READINGS FOR TECHNICAL WRITERS

C. W. Chang

READINGS FOR TECHNICAL WRITERS

DEBRA JOURNET
Louisiana State University

JULIE LEPICK KLING
Texas A&M University

SCOTT, FORESMAN AND COMPANY   GLENVIEW, ILLINOIS
Dallas, Tex.   Oakland, N.J.   Palo Alto, Calif.   Tucker, Ga.   London, England

An Instructor's Manual is available. It may be obtained through a local Scott, Foresman representative or by writing to English Editor, College Division, Scott, Foresman and Company, 1900 East Lake Avenue, Glenview, IL 60025.

Library of Congress Cataloging in Publication Data

Main entry under title:

Readings for technical writers.

Includes index.
1. Readers—Technology. 2. Technology—Addresses, essays, lectures. 3. English language—Technical English. 4. College readers. 5. English language—Rhetoric. I. Journet, Debra. II. Kling, Julie Lepick.
PE1127.T37R42 1984   808'.042'0246   83-16482
ISBN 0-673-15557-9

Acknowledgments

Chapter 1 *Hydrogen*—From *The Condensed Chemical Dictionary* by Gessner G. Hawley. Copyright © 1977 by Van Nostrand Reinhold Company. Reprinted by permission of the publisher. *Cancer*—From *Contrary to Nature* by Michael B. Shimkin. U. S. Government Printing Office, 1975. *Cancer*—From *Encyclopedia and Dictionary of Medicine and Nursing* by Benjamin F. Miller and Claire Brackman Keane. Copyright © 1972 by W. B. Saunders Company. Reprinted by permission. *Terrorists and Terrorism*—Konrad Kellen, "Terrorists—What Are They Like?" A Rand Report, 1979, pp. 6–7. *What Is a Computer?*—From "What Is a Computer?" Reprinted from *Apple* Magazine with the permission of Apple Computer, Inc. *Manufacturers' Sales Representative*—"Manufacturer's Sales Representative," by John C. Warren. U. S. Government Printing Office, 1980. *Crude Oil*—Refinery Siting Workbook, Appendices A & B, Volume 1, U. S. Department of Energy, 1980, pp. 4–5. *Alcohol Problems*—"Alcohol Problems: Patterns and Prevalence in the U. S. Air Force" by J. Michael Polich and Bruce R. Orvis. Copyright © 1979 The Rand Corporation. Reprinted by permission.

Acknowledgements are continued on page 222.

PREFACE

The purpose of *Readings for Technical Writers* is to provide students in technical writing courses with examples of effective professional writing. This text fills a need shared by many technical writing students. Upon beginning a technical writing course, most students have command of a good deal of technical information, but their experience in writing is usually limited to the highly specialized form of the college essay, directed to the equally specialized audience of the college teacher. This kind of writing, however, bears little relation to the writing most students will be asked to produce in professional situations. Thus, the goal of the technical writing course is to introduce students to the many kinds of professional technical writing, and help them produce forms of writing that meet the requirements of particular professional situations. Many students, however, have never seen examples of such forms of writing. *Readings for Technical Writers* responds to that need. The selections in this book furnish students with models that they can imitate and adapt, and that will help them become familiar with the range of technical writing.

Organization

Readings for Technical Writers is divided into five sections: Definition, Description, Instructions, Proposals, and Reports. This order reflects the sequence of assignments made in many technical writing courses, which in turn reflects the topical organization of most technical writing textbooks. It also provides a progression of increasingly complex modes of writing.

The first three sections—Definition, Description, Instructions—exemplify three forms of technical writing that usually have specific and easily definable aims: to explain the meaning of an unfamiliar term, to describe how an object exists in space or a process in time, to offer instructions or guidelines. Because the writer's purpose in these kinds of writing is usually simple, he or she will generally adopt a simple pattern of organization and an uncomplicated format.

The last two sections of *Readings for Technical Writers*—Proposals and Reports—provide examples of writing more complex in structure and purpose. The aims of such writing are many and diverse: to propose research, to obtain funding, to determine feasibility, to survey information, to assess problems, to ascertain causes, to offer solutions. These more complicated forms of writing require the writer to employ more complex patterns of organization and format.

Features

Readings for Technical Writers begins with *To the Student,* a section that discusses the principles of good technical writing. *Introductions* to each chapter present the specific characteristics and considerations of definition, description, instructions, proposals, and reports. This detailed information enables students to evaluate the readings as well as produce examples of their own.

The *readings* provide a variety of writing samples, indicating the considerable diversity of technical writing. The examples are real responses to real professional situations in government, industry, and business. Most importantly, all the readings demonstrate the qualities of good technical writing: clarity, logic, effective organization, concreteness, specificity, and efficiency. To highlight the impact of layout and graphics on technical writing, eight selections have been reproduced as closely to their original appearance as possible.

Each reading is preceded by a *headnote* that describes the context—audience and purpose—in which the selection was written. The headnotes also point out some of the outstanding features of the writing. *Questions for Analysis and Discussion* and *Applications* follow each selection. The questions ask the student to consider how a selection is organized and executed and how that organization and execution are related to the writer's intended audience and purpose. The applications offer suggestions for further writing. A *Glossary* of writing terms at the end of the book gives students a handy reference tool.

Acknowledgments

We thank the many individuals and organizations who have helped in the making of this book. We are grateful to the following for providing us with examples of on-the-job writing:

The Air Transport Association of America

Douglas M. Bonham,
Heath Company

A.G. Brazil,
Douglas Aircraft Company,
McDonnell Douglas Corporation

Claire Caskey,
Department of English,
Clemson University

A. C. "Bud" Clark,
Arizona Public Service

Ken Cook,
Ken Cook Company

Alberta L. Cox,
Naval Weapons Center

Harriet L. Duzet,
IBM

Ronald M. Field,
Westinghouse Electric Corporation

C. Marks Hinton, Jr.,
Underwood, Neuhaus & Company

Richard T. Huber,
U.S. Fish and Wildlife Service

J. Lisko,
Douglas Aircraft Company,
McDonnell Douglas Corporation

Lois McAllister,
Badische Corporation

Jim McKinley,
Celanese Corporation

Robert Miller,
Real Estate Research Corporation

McLane Manufacturing, Inc.

Michael D. Murphy,
Department of Landscape
Architecture,
Texas A&M University

B.G. Niznik,
Bechtel

C.R. Pearsall,
Deere & Company

David E. Ross,
SCS Engineers

Dale Scoville,
Vector Electronics

Nan Booth Simpson,
Southern Living

Margaret L. Somers,
Department of Rhetoric,
University of Minnesota

Vail Associates, Inc.

John R. VanSurdam,
American Enka

B. L. Zoumas,
Hershey Food Corporation

The staff of the Documents Division of the Sterling M. Evans Library at Texas A&M have provided much useful assistance.

We also wish to thank the Departments of English at Clemson University and Texas A&M University and our students in technical writing. We owe a special debt to Professor Norman Grabo, now of the Department of English, University of Tulsa, for his advice and encouragement in the early stages of this project. Thanks are also due to the staff of Scott, Foresman, especially Harriett Prentiss, Amanda Clark, and Kathy Lorden.

Finally, we thank all those who have endured us and endured with us during this project: Alan Journet, Janice Ketcham, Michael Ketcham, Will Kling, Anthony O'Keeffe, and Katherine O'Brien O'Keeffe.

Debra Journet
Julie Lepick Kling

ABOUT THE AUTHORS

Debra Journet, Assistant Professor of English at Louisiana State University, holds a B.A. from Newcomb College of Tulane University and a Ph.D. from McGill University. While a technical writer with the Cyclotron Institute of Texas A&M University, she acted as primary editor and coordinator of the institute's annual report. She also edited professional journal articles, compiled grant proposals, and wrote technical correspondence and memoranda. Professor Journet, the recipient of an NEH fellowship, has also authored several journal articles. Currently, she is the coeditor of the forthcoming *Research in Technical Communication* (1984), a collection of bibliographical essays in the technical writing field.

Julie Lepick Kling, Assistant Professor of English at Texas A&M University, received her A.B. from the University of California, Santa Cruz, and her M.A. and Ph.D. from the State University of New York at Buffalo. As an engineering writer for SCS Engineers, she prepared proposals and reports for private industry as well as municipal and federal agencies. Professor Kling has also worked as a consultant writing computer software documentation. The recipient of an NEH fellowship and several other awards and grants, she is a member of the Society for Technical Communication. She has written numerous articles and reviews, as well as *Opportunities in Computer Science* (1984).

CONTENTS

*Denotes reading selections that are reproduced as close to their original appearance as possible

2

DESCRIPTION   41

3

INSTRUCTIONS   79

4

PROPOSALS   127

5

REPORTS   165

GLOSSARY   217

INDEX   224

TO THE STUDENT

The purpose of *Readings for Technical Writers* is to acquaint you with good examples of the kinds of writing that you may be expected to produce on the job. By *technical writing* we mean writing in the service of science, technology, business, and management.

Two-thirds of the U.S. workforce is currently employed in positions that require handling and communicating large amounts of information.[1] We are experiencing an information revolution, largely because of advances in electronics. These advances have greatly expanded the amount of information that can be collected, stored, and retrieved.

Machines may be able to store and process information, but the use of this information remains the task of the educated individual. And to be useful, information must be communicated to those who need it to develop research, make decisions, and plan for the future. Hence good communication skills are vital for success in our personal and professional lives. Employers are realizing more and more that an employee's specialized knowledge is of little value unless he or she can communicate that knowledge clearly, simply, economically, and (usually) in writing.

In selecting material for this book, we have been guided by two goals: (1) to provide a wide selection of the standard forms of technical writing in a variety of subjects, and (2) to present examples of effective technical writing. We'll examine the various forms of technical writing—definition, description, instructions, proposals, and reports—and the various types of each form in the introductions to each chapter. Before you begin to master these forms, however, let's discuss some of the features that make these selections examples of *good* technical writing, and, in so doing, note a few ways to improve your own writing.

---

[1]"Where Hot New Careers Will Open in the '80's," *U.S. News and World Report,* Dec. 7, 1981, p. 70.

PURPOSE AND AUDIENCE

Before beginning to write, the successful technical writer must have a clear understanding of the purpose and potential audience of the prospective document. In other words, you need to know why you are writing and to whom.

Let us take the relatively simple case in which you wish only to inform your reader, that is, to transmit certain information. Even in this situation, you must answer a number of questions to determine how best to communicate that information:

- What is the reader's interest in or need for this information?
- What does the reader expect to gain from the document? What should he or she be able to know or do after reading it?
- How strong is the reader's motivation to obtain this information?
- Does the reader need to know all there is to know about a subject, or does he or she require only selected information?
- What knowledge of the subject will the reader bring to his or her reading of the document? Expert? Some? None?

These and similar questions demonstrate how closely the purpose of a document is tied to a clear characterization of its potential reader or readers. And very often a single document, such as a report or proposal, will be read by a number of different readers, with varying interests, knowledge, and needs.

Often the purpose of a document is to persuade the reader to take some kind of action or adopt a certain attitude. Actually, any document must be persuasive to the extent that it fosters confidence in the reader that the information transmitted is accurate and reliable. Certain types of technical writing have a more obvious persuasive intent, however. For example, the writer of a grant application or proposal wants to convince the reader to accept and fund the proposed project. Similarly, the writer of a report may wish to convince the reader to accept the report's conclusions and implement its recommendations.

Only after you have developed a detailed idea of your purpose and potential reader should you try to begin writing. At this stage beginning writers often panic, fearing that they have nothing to say. The blank page of paper seems to reflect a sudden utter blankness of thought.

Actually, the problem is usually not that the writer has nothing to say but, rather, just the opposite; too much is going on in the writer's mind—audience, purpose, and the information itself. The problem is deciding how and where to begin. To help you decide, develop a plan of organization, no matter how brief. Once you have a plan on paper, you have something to guide your thoughts. You may find it easier to organize your writing if you think of it in terms of two aspects: content and expression.

CONTENT

Content refers to the information that you wish to communicate. It also includes your purpose and your reader, because the information you transmit is shaped by those essential considerations of audience and intent.

Most professional writers gain control of their information by jotting down the major points they wish to make or topics they want to cover. These notes more closely resemble a shopping list than they do a formal outline. In fact, most seasoned writers do not work from a formal outline, nor do they proceed in an ordered sequence from beginning to end. On the contrary, once a writer has an informal list of the items to be covered, he or she may be ready to develop something that will end up in the middle or even at the end of the finished document. Moreover, it's not always practical to begin at the beginning. For instance, if you are writing the introductory summary of a report, you can't very well introduce or summarize anything until you've written the report itself.

Think of writing as a process of assembling: Working from an informal list of topics, you will write small sections of a document at a time and make frequent use of the writer's most valuable tools (after the dictionary)—scissors and tape. If you are composing on a word processor, you can move sentences and paragraphs around at will. Keep in mind that a document is *composed*—literally, "put together."

EXPRESSION

Expression is the realization of a document's content, the written or printed form of information aimed at a particular audience for a well defined purpose. Expression encompasses both the visual and verbal features of a document. The visual features of a text include its format, presentation, and use of graphics. A text's verbal features are those qualities that make good writing readable, including interesting sentence structure, adequate transitions, effective word choice, and (that elusive notion) style.

The Visual Features

Format and Presentation
In this book we aim to present a wide selection of the formats used to communicate information in government, business, industry, and research. By *format* we mean how words and graphics are arranged on a page and how the content is divided into sections—in other words, *layout* and *organization*.

There are general formats for certain kinds of writing, such as memoranda, business letters, proposals, progress reports, and manuals. But often the general format must be adapted or modified to suit the needs of the particular audience and purpose of the document. Frequently, the client for whom a document is prepared will provide detailed specifications of an acceptable format. Other kinds of technical writing may allow the writer a wider degree of choice concerning content and presentation.

Although the beginning technical writer may not like being told "this is how you must prepare this document," format requirements can serve an important purpose. By standardizing documents of a similar sort, format requirements help the reader obtain the information contained in a document as quickly and easily as possible. Once mastered, formats make the task easier for the writer, too, by supplying guidelines that reduce the amount of time and energy spent on determining how to present the information.

As a general rule, *presentation*—how a document looks—contributes greatly to the reader's attitude. An attractive document will be well received, as the reader is more likely to trust the authority of the information if the document has been prepared with care and attention. And certain features of layout can make the writing easier to read. Headings, lists, and the generous use of white space emphasize important information, making the page less cluttered and easier on the eye. Good design, attractive layout, a legible typeface, high quality reproduction, and, above all, thorough proofreading contribute to the overall effectiveness of all technical writing. Even when you're penning only a handwritten memo to your boss, if he or she has trouble deciphering your careless scrawl, you haven't communicated effectively. ■

Graphics

Most technical writing relies, in part, on graphics (or visual aids) to help the reader understand information not readily presentable in writing, or to interpret the significance of a large amount of (usually quantitative or numerical) data. Visual aids can also increase the reader's interest in a topic, particularly when it is unfamiliar.

In formal documents, visual aids are usually classified into two categories: *tables* and *figures.* Tables arrange verbal or numerical data into rows and columns. All other visual aids, such as charts, graphs, diagrams, line drawings, photographs, or maps, are considered figures. Less formal types of technical writing, such as brochures or product support literature, frequently use attractive illustrations as a persuasive tool—to convince the reader to accept a particular point of view, purchase a product, or take some other action.

The Verbal Features

Readability

One of the writer's most important tasks is to guarantee that the writing is readable. The *readability* of a document is a measure of the ease with which it can be read and comprehended. Thus, readability involves adapting the language of a document to its purpose and audience. In short, a document is highly readable if its language meets the needs of its readers and fulfills the purpose for which it was written.

A common measure of the readability of a piece of writing is the number of syllables in its words, words in its sentences, and sentences in its paragraphs. The length of words, sentences, and paragraphs affects the complexity of the writing. Readers with less developed reading skills will need shorter words, sentences, and paragraphs if the writing is to be understood. More literate readers can comprehend more complicated writing.

But readability depends on more than simply the reading skills of the audience. It also involves adapting the way the subject is treated to fit the reader's needs, interests, and level of knowledge. Many complex formulae have been developed that compute readability by measuring the number of words, sentences, or even syntactical units. While these are useful, the readability of a document necessarily encompasses many other factors, including the following aspects of all good writing.

Sentence Structure (or Syntax)

The normal sentence in English is arranged in a pattern consisting of subject-verb-complement (SVC) or subject-verb-object (SVO). "John is tall" or "John hit the ball" are simple examples of this pattern.

To express more complicated ideas and relations between these ideas, sentences grow more complex and rely on varied structures. *Coordination,* making ideas of equal importance grammatically equal in the sentence, and *subordination,* making less important ideas grammatically dependent on more important ideas, are two ways to achieve variety. *Parallelism* is an effective way to balance related ideas by using similarly structured words, phrases, or clauses: "The samples were collected, freeze-dried, and measured." *Parenthesis*—setting off ideas from the rest of the sentence by dashes, commas, or parentheses (as in this example)—or *ellipsis*—omitting words implied by the context of the sentence—can also make your writing more economical and interesting. And you can also achieve variety and create emphasis by inverting the normal order of the sentence: "Down the stairs John tumbled, hitting his elbow as he fell."

Finally, technical writing is usually most effective in the *active voice.* With the active voice, the actor or doer of the action is the subject of the sentence: "John hit the ball." With the *passive voice,* the object or recipient of the action becomes the subject of the sentence: "The ball was hit by John." The passive voice has some legitimate uses—for example, when the subject is unknown or unimportant: "The samples were collected, freeze-dried, and measured." But the passive voice requires more words than does the active and usually leads the writer into awkward constructions. Moreover, the active voice is more emphatic and idiomatic.

Transitions

Transitional words and phrases also increase a document's readability. *Transitions (for example, therefore, however, first, then,* and so on) are signposts that guide the reader through the writing, highlighting major points, explaining the relationship of statements, indicating organization, pointing to supporting evidence—in general, telling the reader how to progress from one sentence or paragraph to the next. Think what would happen to the meaning of a complex mathematical equation if all the relational symbols were eliminated. Writing that lacks effective transitional words and phrases is no less confusing.

Word Choice (or Diction)

Good technical writing requires that you carefully consider the words you use. A long and complicated sentence made up of short, one- or two-syllable words may be much more readable than a brief sentence using more difficult terms. In suiting your vocabulary to your reader, you must assess his or her knowledge of your subject. Obviously, you would use very different words to explain the conversion of food to energy in the human body when addressing a class of grade school children, a group of high school biology students, or a gathering of medical experts in metabolic dysfunctions. Even when writing for experts, use only those specialized terms that are essential for clear communication in your field. Avoid jargon and needlessly technical words that will mystify or confuse your reader. Prefer the familiar word to the unfamiliar, and if you must use an unfamiliar term, provide a short definition. Remember, the reader often needs to extract information from a document as quickly as possible and will not always have time to consult a dictionary to find out what a word means.

Style

In spoken communication, the meaning of a statement results, in part, from the speaker's tone and inflection. Written language also has a "voice": its style. *Style* is the accumulation of all the choices a writer makes. Each of us has a peculiar style of writing, as individual as our speaking voice. Although technical writing is sometimes called impersonal, it too reflects the style of its writer, though in less obvious ways than do other forms of writing.

Technical writing is often assumed necessarily to be dull writing. But technical writing's reputation for dullness stems, in large part, from the quantity of bad technical writing—obscure and jargon-ridden—that abounds. Good technical writing shares the virtues of all good writing: clarity, unity, coherence. Although much technical writing, including some of the readings in this book, demands a simple, terse, and uncluttered style, technical writing can also be elegant, sophisticated, and stylistically interesting, as many other selections in this book demonstrate. Again, the stylistic choices you make depend on the audience and purpose of your writing.

REVISION

Once you have composed your document, your task is not over. To a large extent, good writing is rewriting. Give yourself some time, then go back to your work with a critical eye or ask another reader to evaluate it. Is the organization logical? The phrasing clear? The punctuation and spelling correct? The style appropriate? Most successful writing goes through several drafts. Rewriting, or editing, your work is as important as writing it for the first time.

Much of the writing you may do on the job will be completely your own: memoranda, letters, short reports on narrowly defined topics. However, most longer documents, such as manuals, proposals, and full-scale reports, will be written by teams of individuals working together, each bringing to the project his or her own specialized knowledge. All the observations we have made concerning purpose, audience, content, and expression apply to your work as a member of a writing team. In addition, when working on a team, remember not to become too protective of your own work. Remember that in the development of a document any contribution can be altered or even deleted in the interests of the larger purpose of the project.

1

DEFINITION

To *define* **a term is to set** its limits, to draw a semantic boundary that sets it off from all other terms. Without such definition, language would be a meaningless babble and understanding an impossibility. In everyday experience, how many times have you been asked to "define your terms," to "say what you mean"? Technical writing, in particular, aims at precise encoding of information. Without careful specification of terms, information can be misunderstood; the resulting losses—of time, money, and efficiency—can be considerable.

Different types of definition are required for different terms and in different contexts. We are all familiar with the standard dictionary definition of words; often, however, these definitions are technically incomplete, inexact, or even out-of-date. A more rigorous form is the *logical definition,* also known as the *formal* or *sentence definition.* The logical definition has three parts: the *term* to be defined; the *class* to which the term belongs; and the *differentiae,* or distinguishing characteristics, that set the term off from other members of the same class. For instance, in the selection on pages 27–28, John C. Warren defines *manufacturers' sales representative* as "a professional salesman who is in business for himself." Here, manufacturers' sales representatives are compared with other members of the general class of professional salesmen but are also distinguished from those other members according to who employs them: manufacturers' sales representatives do not work for companies or organizations; they work for themselves.

All the definitions in the following selections are *extended* (or *expanded*) *definitions.* Almost all use logical definition; in addition, they employ a variety of other techniques to define the subject adequately for the needs of a particular project or situation. These techniques include:

- *analogy:* likening the term being defined to a more familiar term, object, or experience.

- *analysis and classification:* separating the term into parts in order to explore the relation of part to part and part to whole.
- *cause and effect:* describing how a term developed or explaining the causes and effects of the phenomenon to which the term refers.
- *comparison and contrast:* explaining the similarities and differences between one phenomenon and another.
- *description:* indicating what the phenomenon defined looks like or explaining the process or sequence of activities through which it moves.
- *etymology:* revealing the historical roots of a word to help elucidate its current meaning.
- *exemplification:* providing specific examples to make the term concrete.
- *explication:* explaining the meaning of key words in the logical definition.
- *figurative language:* using vivid, dramatic, or forceful expressions not in their literal sense, but to reveal unexpected connections or associations; common figurative devices include *metaphor, simile,* and *personification.*
- *history of the subject:* placing the term in the context of its discovery or development.
- *illustration:* using graphic aids to assist in producing a fuller understanding of the term.
- *negation or elimination:* specifying what the phenomenon defined is *not* in order to clarify what it is.
- *operational definition:* directing the reader to a time or place where the phenomenon, process, or object to which the term refers can be observed; describing the conditions under which the phenomenon defined comes into being.
- *stipulatory definition:* specifying or limiting the meaning of a term for the purposes of a particular occasion.

Contrary to what we might expect, words do not always have a single, stable meaning. The connection between a word and its meaning is a matter of conventional association and sometimes even of simple stipulation. This arbitrary connection between words and meanings is illustrated by *neologism,* the "new word" made up to fit an as yet unnamed phenomenon. Such new words are especially common in technical and scientific writing: *robot, pasteurization, quark, videotape,* and *cyclotron* are all words created rather arbitrarily to fit various new objects and procedures. Even in ordinary speech, the meanings of words change with time. For instance, according to the *Oxford English Dictionary,* the word *engine* originally meant "wit, skill, or native talent" (compare the related word, *ingenious*); today *engine* designates a mechanical device.

Some terms do have a single and unchanging meaning. This is especially true of many scientific terms. The names of the elements refer to certain absolute atomic and molecular properties; for instance, the term *hydrogen* will always refer to a particular molecular arrangement. Other terms, often no less technical, may change their meaning not only in time but also in different contexts. For instance, the term *information,* as we use it here, refers to verbally encoded facts and conclusions; to a specialist in information sciences, however, *information* has a far more narrowly specified meaning. Some terms, indeed, must be operationally defined by the conditions that define their very existence, as John Warren defines a *manufacturers' sales representative.* Moreover, the scope, focus, or goal of a particular project, problem, or course of research may require that a term be defined with a special purpose or from a restricted perspective, as in Konrad Kellen's definition of *terrorism* (see pp. 20–21). Thus, the definition of a specific term in its immediate context is rarely the final word on the subject.

The formulation of good definitions is one of the most demanding aspects of technical writing, as definitions form an essential component of descriptions, manuals, proposals, and reports. The definitions included in this chapter come from many kinds of technical writing, are directed at a range of readers, and deal with a variety of subjects. Moreover, these definitions illustrate how the complexity and scope of the definition, like that of all technical writing, is dictated by both the writer's purpose and the intended reader's abilities.

Hydrogen

Many scientific and technical terms have an agreed-upon meaning that does not shift according to the context in which the term is used. In such cases, scientists establish certain indisputable physical properties that distinguish this phenomenon from any other. The periodic table of elements is a familiar example. The following definition of *hydrogen,* from *The Condensed Chemical Dictionary,* begins by establishing the properties of hydrogen and then proceeds to a more general discussion of the element.

1   **hydrogen H.** Element of atomic number 1. Group 1A of Periodic Table. Atomic weight 1.008; valence I. Isotopes: deuterium (H^2), tritium (H^3). Discovered by Cavendish in 1766; named by Lavoisier in 1783 (water-maker).

2   Properties: A diatomic gas; density 0.0899 g/l; sp. gr. 0.0694 (air = 1.0); specific volume 193 cu ft/lb (70°F); frz. pt. −259°C; b.p. −252°C. Autoignition temp. 1075°F. Very slightly soluble in water, alcohol and ether. Noncorrosive. Can exist in crystalline state at from 4 to 1 degree Kelvin. Classed as an asphyxiant gas.

3   Occurrence: Chiefly in combined form (water, hydrocarbons, and other organic compounds); traces in earth's atmosphere. Unlimited quantities in sun and stars. It is the most abundant element in the universe.

4   Derivation: (1) Reaction of steam with natural gas (steam reforming) and subsequent purification; (2) partial oxidation of hydrocarbons to carbon monoxide, and interaction of carbon monoxide and steam; (4) dissociation of ammonia; (5) thermal or catalytic decomposition of hydrocarbon gases (see "Hypro" process); (6) catalytic reforming of petroleum; (7) reaction of iron and steam; (8) catalytic reaction of methanol and steam; (9) electrolysis of water; (10) Thermochemical decomposition of water is under consideration as a large-scale source of hydrogen; it is said to involve addition to water of calcium, mercury, and bromine and temperatures between 500 and 800°C (Euratom plan). The chief industrial method is (1); but (7) is being developed in connection with hydrogasification, and (9) has long-range future possibilities by use of a special plastic electrolyte originally developed for fuel cells.

5   Method of purification: By scrubbing with various solutions (see, especially, the Girbitol absorption process). For very pure hydrogen, by diffusion through palladium.

6   Grades: Technical; pure, from an electrolytic grade of 99.8% to ultra-pure, with less than 10 ppm impurities. See also para-hydrogen.

7   Containers: Steel cylinders; tank cars (in cylinders); pipeline.

8   Hazard: Highly flammable and explosive; dangerous when exposed to heat or flame. Explosive limits in air 4 to 75% by volume.

9   Uses (Gas): Production of ammonia and methanol; hydrocracking, hydroforming and hydrofining of petroleum; hydrogenation of vegetable oils; hydrogenolysis of coal; reducing agent for organic synthesis and metallic ores; reducing atmospheres to prevent oxidation; as oxyhydrogen flame for high temperatures; atomic-hydrogen welding; instrument-carrying balloons; making hydrochloric and hydrobromic acids; production of high-purity metals.

10   (Liquid): Coolant and propellant; fuel for nuclear rocket engines for hypersonic transport (mach 6); missile fuel; cryogenic research.

11   Shipping regulations: (gas) (Rail) Red Gas label; (Air) Flammable Gas label. Not accepted on passenger planes. (liquid) (Rail) Red Gas label; not accepted by express. (Air) Not accepted. For other carriers, consult authorities.

12   *Note:* In view of the energy-releasing capacity of hydrogen, especially its isotope tritium (H^3), which occurs in thermonuclear reactions, hydrogen is regarded by some authorities as the ultimate energy source. Splitting of the water molecule by thermochemical or nuclear reactor technology is considered to be an active future possibility for producing hydrogen in high volume for use as an energy source. The electrolytic method is too inefficient for this purpose. Research on development of a controlled hydrogen fusion reaction is in progress, but positive results are unlikely in this century. The use of hydrogen as a transportation fuel for naval ships and aircraft is under active investigation.

Questions for Analysis and Discussion

1. Although this definition does not say everything there is to say about hydrogen, most of what it contains would be generally accepted by all scientists. How is this different from the other definitions in the chapter?

2. Why has the Note been appended to the definition rather than integrated within it? What are the differences between the Note and the rest of the definition?

3. Although the characteristics that define hydrogen are no longer subject to change or scientific debate, this particular definition contains features that will eventually make it out-of-date. Identify some of these features. How do they help suggest the distinction between science and technology?

4. Select a term of basic importance in your field of study. Compare the way this term is defined by a standard English language dictionary (such as *Webster's New Collegiate Dictionary* or the *American Heritage Dictionary*), a general purpose encyclopedia (such as the *Encyclopaedia Britannica*), and a special purpose dictionary or handbook for experts in your field.

Application

Select another term in your field and write a brief but unambiguous definition in a form suitable for a reference work. Assume you are writing to someone whose general knowledge of the field is similar to your own.

Michael B. Shimkin

Cancer

U.S. government agencies prepare many kinds of publications for a variety of audiences. *Contrary to Nature,* from which this definition is excerpted, was issued by the Department of Health, Education and Welfare (now the Department of Health and Human Services) to provide a general commentary on the history of knowledge concerning cancer. Because the definition was written for an audience with no medical training, it assumes little prior knowledge, explains unfamiliar terms, and remains at a fairly high level of generality.

1   Cancer is a word in the English language, derived from the Greek word for crab, Karkinos. Among its many synonyms are malignant tumor and malignant neoplasm (from the Greek for new growth). Subgroups of cancer, describing the body tissues of origin, include carcinoma, sarcoma, melanoma, lymphoma and many other related or combined terms. The systematic classification of cancers, based upon microscopic appearance, replaced older, more colorful names such as scirrhus (hard, or a scar), and nolimetangere (do not touch me).

2   Cancer is a word that stands for a great group of diseases that afflicts man and animals. Cancer can arise in any organ or tissue of which the body is composed. Its main characteristics include a seemingly unrestricted or uncontrolled growth of abnormal body cells, with the resultant mass compressing, invading and destroying contiguous normal tissue. Cancer cells then break off or leave the original mass and are carried by the blood or lymph to distant sites of the body. There they set up secondary colonies, or metastases, further invading and destroying the organs that are involved.

3   Cancer kills. Unless it can be successfully treated, cancer kills inexorably, slowly, and unpleasantly. Death is the basic fact about cancer. That is what makes cancer a dread disease. All other features of cancer are secondary to its deadliness.

4   Two important characteristics of cancer are anaplasia and autonomy. Anaplasia is the loss of normal appearance under the microscope, with the cells comprising the cancer being disorganized in arrangement and varying in size and shape. Autonomy is the loss of inhibition of cell growth, with resultant

semi-independent behavior and function. The diseases grouped under cancer are now second only to diseases of the heart and blood vessels as killers of the people of the Western world. In the United States of 1970, among its 210 million people, approximately 660,000 developed cancer during the year, and approximately 330,000 died of cancer during the year.

5   There has been a steady, striking increase in the number of cancer deaths in the United States and in Europe during the past century. Some of this rise is due to better diagnosis of cancer and better reporting of cancer cases. But a much more important factor has been the increasingly older population, saved from earlier death from infectious diseases and other causes. Although cancer does not spare the young, its main impact is reserved for the elderly. This is perhaps the only merciful aspect of cancer.

Questions for Analysis and Discussion

1. What features of this definition indicate that it has been written for a nonspecialist reader?

2. This definition supplies a good deal of information about the history or etymology of the terms associated with cancer. What purpose might such information serve?

3. Why do you think the writer has chosen to explicate the meanings of some terms in his definition (for example, *anaplasia* and *autonomy*) and not the meanings of others (for example, *sarcoma, melanoma, lymphoma*)? Do you find this inconsistent, or is it a legitimate consideration of the reader's need for information?

4. This definition of cancer incorporates affective or emotional features that appeal to the reader's feelings but that would be excluded from rigorously objective writing. Identify these features and discuss their effects. Compare this appeal to the reader's emotions to the more objective, clinical treatment of cancer in the following definition. Which do you find more effective in convincing the reader of the seriousness of the disease?

Application

Rewrite the definition you composed in the previous Application (p. 14) so that it is appropriate for a lay reader. How do the two definitions differ in their content and expression? What specific techniques did you use to adapt your specialized material for a general audience?

Benjamin F. Miller
and Claire Brackman Keane

Cancer

The following definition of cancer comes from the *Encyclopedia and Dictionary of Medicine, Nursing, and Allied Health,* which aims to provide an authoritative vocabulary for students in nursing and paramedical sciences. This purpose and audience have determined the context for the discussion of cancer. Because the definition is directed to students in medical fields, it explains the disease in terms of its diagnosis and treatment and deals with it as a specifically medical phenomenon, rather than as a biological or pathological entity.

1   **cancer** (kan'ser) any malignant, cellular tumor. adj., can'cerous. Cancer is a group of neoplastic diseases, in which there is new growth of abnormal cells. Normally the cells that compose body tissues grow in response to a normal stimulus. Worn-out body cells are regularly replaced by new cell growth which stops when the cells are replaced; new cells form to repair tissue damage and stop forming when healing is complete. Why they stop forming is unknown, but clearly the body in its normal processes regulates cell growth. In cancer, cell growth is unregulated. The cells continue to reproduce until they form a mass of tissue known as a tumor. Not all tumors are malignant; those which are noncancerous are referred to as benign tumors. Benign tumors vary in size, and may grow so large that they obstruct organs or cause ulceration and bleeding. They are encapsulated, do not metastasize, and usually can be removed by surgery without difficulty.

2   Malignant tumors grow in a disorganized fashion, interrupting body functions and robbing normal cells of their food and blood supply. The malignant cells may spread to other parts of the body by (1) direct extension into adjacent tissue, (2) permeation along lymphatic vessels, (3) traveling in the lymph stream to the lymph nodes, (4) entering the blood circulation, and (5) invasion of a body cavity by diffusion.

3   *Causes.*   Cancer is many different diseases and no one factor can be pinpointed as the cause of the various types of malignant growths. Environmental, hereditary, and biological factors are all known to play a role in the development of cancer.

4   Environmental causes are believed to account for at least 50 per cent and perhaps, in some types, as much as 80 per cent of all cancers. For example, cigarette smoking is directly related to approximately 90 per cent of all cancers of the lung. Other environmental carcinogens include industrial pollutants and radiation. Among the chemical carcinogens are arsenic from mining and smelting industries; asbestos from insulation, at construction sites and power plants; benzene from oil refineries, solvents, and insecticides; and products from coal combustion in steel and petrochemical industries. Each year new products that in all probability are carcinogenic are being produced by industrial operations. A major concern is the occupational and environmental hazards these chemicals present to those who work in or live near these plants.

5   Radiation from prolonged exposure to the ultraviolet rays from the sun and from injudicious use of diagnostic and therapeutic procedures involving x-rays and radioactive substances is also a significant factor in the incidence of cancer, particularly in the development of cancer of the skin, bone marrow, and thyroid.

6   Hormones, especially the synthetic estrogens given to forestall the effects of menopause and to prevent spontaneous abortion, are directly related to some cancers of the female reproductive organs.

7   Viruses as causal agents in the development of cancers have been subjected to intensive research efforts in recent years, and, while a number of cancers can be produced in experimental laboratory animals, there is still no irrefutable evidence that cancers in humans are caused by viruses. An exception may be the Epstein-Barr virus, which may have a causal association with Burkitt's lymphoma and certain cases of nasopharyngeal cancer. There remains, however, the intriguing fact that viruses are capable of introducing new genetic material into a normal cell and transforming it into a malignant one, and that cell reproduction may be altered when viruses interact with such carcinogens as chemicals and radiation. It is not known exactly how these properties enhance the ability of malignant cells to thrive under adverse conditions and to metastasize to other parts of the body and produce another cancerous tumor. Recent studies have shown that an extracellular enzyme plays an important role in the transmission of genetic information to the cell and thereby facilitates the reproduction of cancer cells. The enzyme is called reverse transcriptase because it reverses the usual mechanism for replication of genetic information; that is, whereas in normal cellular replication the DNA is the template for RNA copies, in the presence of the enzyme, RNA serves as the template for DNA copies.

8   The incidence of cancer in certain populations suggests that other factors are important in its development. It is known, for example, that some families show a high incidence of malignancy among [their] members, but there is no definite hereditary pattern. There also is a high incidence of cancer in persons receiving drugs for immunosuppression, yet cancer itself is immunosuppressive. It is suggested that prolonged suppression of the body's immune response may eventually impair its ability to distinguish between self and nonself and thus render it unable to destroy malignant cells. When cancer

itself acts to suppress the immune response, it may be the result of an overwhelming demand on the body to destroy more foreign cells than it is prepared to cope with at any given time.

9 *Classification.* Cancers are classified on the basis of two factors: the type of tissue and the type of cell in which they arise. Using this classification system, it is possible to identify over 150 types of cancer in humans. In the classification of cancers according to the type of tissue from which they evolve, there are two main groups: SARCOMAS and CARCINOMAS. Sarcomas are of mesenchymal origin and affect such tissues as the bones and muscles. They tend to grow rapidly and to be very destructive. The carcinomas are of epithelial origin and make up the great majority of the glandular cancers and cancers of the breast, stomach, uterus, skin, and tongue.

10 Cell type affects the appearance, rate of growth, and degree of malignancy. Thus, classification of tumors according to the type of cell from which they are derived is important in deciding the course of treatment for a specific malignancy.

Questions for Analysis and Discussion

1. Both this and the preceding definition of cancer are introductory definitions aimed at readers without extensive specialized medical knowledge. Their assumptions regarding reader comprehension, however, are quite different. In what ways are these different assumptions manifested in tone, style, sentence structure, and word choice?

2. In paragraph 3, the authors suggest that the causes of cancer are hard to pinpoint. How do they organize the complex and rather diffuse subject of the causes of cancer in order to deal with it succinctly and rationally? What devices do the authors use to help the reader follow their pattern of organization?

3. Paragraph 2 is composed of only two sentences. How have the writers employed effective sentence structure to keep these two sentences clear and intelligible?

Application

This definition of cancer assumes some specialized knowledge. Rewrite the last paragraph in the section on causes (paragraph 8) so that it would be understood by a reader who feels more at home with Shimkin's definition of cancer (pp.15–16).

Konrad Kellen
Rand Corporation

Terrorists and Terrorism

Words used in everyday speech may have many different meanings. But when a word is used within the context of a particular project, its meaning must be stipulated if the term is to be of value. The following definition comes from a short report written by Konrad Kellen of the Rand Corporation, a California-based research organization. The report was prepared for Sandia Laboratories, a subsidiary of AT&T devoted to basic and applied research in such areas as aerospace science and energy alternatives. These are areas of research critical to national defense and thus particularly vulnerable to terrorist attack. Kellen, in his definition of *terrorism*, addresses the need to restrict the meaning of this familiar and emotionally charged word in order to suit the purposes of his report and the needs of his readers.

What Do We Mean by Terrorists and Terrorism?

1   Many definitions of terrorists and terrorism have been presented, and just as many (usually valid) objections have been raised against these definitions. Like many other phenomena, terrorists, so mercurial and elusive in real life, are not easily bound in words either.

2   Against definitions that are primarily negative or pejorative, the objection has been raised that the early American revolutionaries would have to be regarded as terrorists by contemporary standards. Thomas Jefferson, who said, "The tree of liberty must be fertilized from time to time with the blood of tyrants," might qualify for the label. Similarly, the men who tried to assassinate Hitler or the men who succeeded in killing his chief representative Reinhard Heydrich in subdued and tortured Czechoslovakia would be terrorists to some.

3   But to put Jefferson or anti-Nazi heroes within the same semantic confines as the perpetrators of the Lod Airport massacre or the murder of the American ambassador in Khartoum would only attest to the uselessness of any definition so wide as to include such disparate elements. It is therefore necessary to introduce some arbitrary criteria for the sake of discussion. The term *terrorism* as used in this Note assumes the following restrictions:

1. Terrorism refers to *contemporary* activity. Historical parallels, even of such recent date as World War II, Korea, and Vietnam, may be illuminating but they are not "the same" as what we understand by terrorism today.
2. Terrorism is distinguished from terror, which is the rule by force and fear "from the top," i.e., by a dictatorial regime.
3. Terrorism is violent action, especially against human beings, or against symbolic targets.
4. Mere threats of violence are not terrorism, *unless* they emanate from a group that has already engaged in terrorist acts.
5. Terrorism is the work of small groups.
6. A terrorist group may or may not have an active working relationship with another terrorist group.
7. A terrorist group must have a political objective, even if it has other objectives as well, e.g., religious objectives.
8. A terrorist act, contrary to a "common" criminal act, must point beyond itself, i.e., the task is not completed with the execution of the act.
9. A terrorist act must instill fear by being violent, visible, irrational, repeatable, or a combination of these.
10. A terrorist act must be extortionist in nature, even if the extortion is not specifically stated at every turn. The equation of terrorism is:

    Violent act committed = More violent acts can be expected, unless or until certain things are done (or discontinued).

4   In a word, terrorism *is* extortion, over time, successful or not, by small groups against large groups.

Questions for Analysis and Discussion

1. What is the effect of Kellen's announcement in paragraph 1 that the term *terrorism* resists definition? How do the specific historical examples he uses help limit this "elusive and mercurial" term?

2. The introduction to this definition (paragraphs 1–3) comprises almost half the definition and serves mainly to explain what terrorism is not. To what extent is this an effective way to introduce a definition? What might be Kellen's purpose in using a disproportionately long introduction?

3. What is there in the nature of this definition that makes a list of qualities more appropriate than an extended discussion? These qualities of terrorism are simply stated; no attempt is made to explain or justify them. Do you find such an approach arbitrary? In what ways is it appropriate to a stipulatory definition?

4. Compare the problems of defining a term such as *terrorism,* which, as Kellen points out, can have different meanings for different persons, with the far more straightforward task of defining a physical entity such as *hydrogen* (pp. 12–13).

Application

Many elements of this definition clearly reflect its time, place, purpose, and point of view. Examine these elements and discuss the way their presence limits the universality and objectivity of the discussion. To what extent is this definition biased? Is such bias legitimate, given the audience and purpose of the document? In one or two paragraphs, write a more general definition of *terrorism* that is as universal and objective as you can make it. What are the differences between your definition and Kellen's? For what purposes might each be more appropriate?

Gene Carter
Apple Computer Inc.

What Is a Computer?

Computers are now widely used in both technological and nontechnological fields. Yet many people, although increasingly dependent on the computer in their daily lives, remain baffled by computer terminology. The following definition was written for such individuals by Gene Carter of Apple Computer Inc. It appeared in his article, "A Consumer's Guide to Personal Computers," in *Apple,* the company's corporate publication. Carter has taken great care to allay the fear that computers are mysterious and incomprehensible. Particularly important in making his subject accessible to his readers is Carter's use of example, history, and analysis.

Also important in helping the reader follow the article is Carter's logical pattern of organization and his thorough introduction to the material. An effective introduction generally indicates the context, purpose, organization, and intended audience of a piece of writing. Sometimes these elements are explicitly stated; sometimes they are implied or indicated by the title or headings. Given the general audience of this article, Carter's logical introduction is particularly useful.

A CONSUMER'S GUIDE TO PERSONAL COMPUTERS

BY GENE CARTER
APPLE COMPUTER INC.

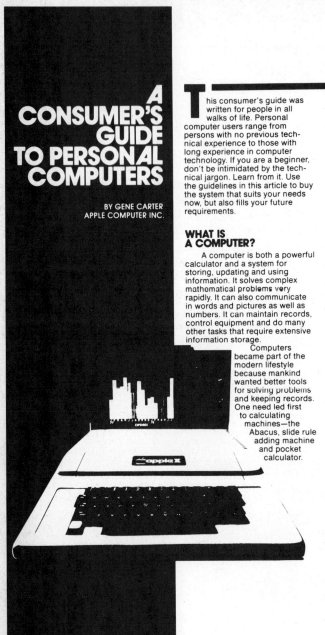

This consumer's guide was written for people in all walks of life. Personal computer users range from persons with no previous technical experience to those with long experience in computer technology. If you are a beginner, don't be intimidated by the technical jargon. Learn from it. Use the guidelines in this article to buy the system that suits your needs now, but also fills your future requirements.

WHAT IS A COMPUTER?

A computer is both a powerful calculator and a system for storing, updating and using information. It solves complex mathematical problems very rapidly. It can also communicate in words and pictures as well as numbers. It can maintain records, control equipment and do many other tasks that require extensive information storage.

Computers became part of the modern lifestyle because mankind wanted better tools for solving problems and keeping records. One need led first to calculating machines—the Abacus, slide rule adding machine and pocket calculator.

The other can be traced back even further through filing cabinets to the picture writing of early civilizations and the notched sticks of primitive man.

The first electronic computer was really a giant calculator named ENIAC (Electronic Numeral Integrator and Calculator). Built in the 1940's with 18,000 vacuum tubes, it filled a large room, consumed 130,000 watts of power, and could be used only by a few experts.

Today's personal computer built with a microprocessor has about five times the computation power of ENIAC, is portable, about 18-inches square, needs about 50 watts of power and can easily be used by all members of a family.

Computers, like some calculators, are programmed—given a series of instructions by the user—to govern their operation. All computers consist of five basic parts. These sub-systems are:

1. CPU or Central Processing Unit

The CPU is the "brain" that manipulates all information and performs all calculations.

2. Control Unit

The CPU is controlled by two kinds of programs. "Software" programs are entered by the machine operator, stored in the memory unit and can be changed as often as desired. "Firmware" programs are built into the system, usually in Read Only Memory (ROM) devices that store instructions permanently. Each instruction generally requires several computer operations. The control unit and firmware enable the computer to perform these operations in the right order and at the proper times.

3. Memory Unit

Software programs and data being processed are stored in the memory unit. This memory can be "randomly addressed," which allows the CPU to store and fetch (write and read) data rapidly. For the computer to handle more and more tasks, Random Access Memory (RAM) must be expandable and able to "swap" programs and information with storage peripherals such as tape cassettes and magnetic disk memories.

4. Input Interface Unit

Information, control signals, and software enter the computer through this unit, which is attached to a keyboard and other peripherals such as tape cassettes and magnetic disk memories.

5. Output Interface Unit

Results of calculations and processed information go out through this interface to a TV screen, printer, tape cassette, or

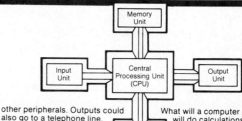
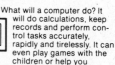

other peripherals. Outputs could also go to a telephone line, home security and environmental system controls, or even to appliance controls.

How does a programmer communicate with a computer? Think of the computer as another country with its own language. If you went to France for a vacation, you could learn French first or hire an interpreter. A computer programmer can learn to use special "machine languages" or he can use interpreter languages that the computer, itself, translates into instructions. Such interpreters include BASIC, FORTRAN, COBOL, PASCAL, APL., etc. Because BASIC is the most versatile and simplest to learn, it is the preferred language for personal computers.

What will a computer do? It will do calculations, keep records and perform control tasks accurately, rapidly and tirelessly. It can even play games with the children or help you compose and play electronic music. A personal computer is a servant that can save time, educate, entertain, control, and guard your home 24 hours a day. It's used by adults for household management, business and technical work, and enjoyment. It's used by children to learn about computers and to develop mental skills and manual dexterity.

Enjoyment is a major reason for the popularity of programmable computers. You don't need to be a programmer to use them; program libraries are available, and many users also get great satisfaction from developing their own personal programs. Some have even taught their computers to operate with voice commands spoken in plain English!

A good personal computer comes ready to use with a built-in keyboard for input and a program library on tape cassettes. It is connected to an audio cassette recorder for program loading and information storage, and to a TV set for output, as easily as speakers are connected to a hi-fi system. Other peripherals are also easy to connect and the best ones have programming aids that help you become an expert at your own pace.

Questions for Analysis and Discussion

1. Identify the elements of an effective introduction—indications of context, purpose, organization, and audience—as they are used in this article. Which are stated directly? Which are implied? Are any omitted or replaced by other elements of the article?

2. How have certain devices, such as graphics, layout, use of historical and everyday examples, been used to make this document less forbiddingly technical and more appealing to the novice reader?

3. What features suggest that the article aims to promote a product as well as inform its readers?

4. Computer terminology is often assumed to be mysterious and esoteric. To what degree is Carter able to avoid the charge of using baffling jargon? What efforts has he made to help the reader understand the specialized terms he uses? Does Carter leave any terms undefined that you feel need further explanation?

Application

Write an extended definition of a term in your field or rewrite your definition for *cancer* (Application, p. 19) so that the layout of your work will enhance its appeal to a general reader. Consider using headings, white space, colors, graphics, and any other features that will make your writing visually attractive. Use the format of the definition to persuade the general reader that your material is interesting and comprehensible.

John C. Warren
Small Business Administration

Manufacturers' Sales Representative

This definition of *manufacturers' sales representative* forms part of the introduction to a bibliography of readings in the field. Issued by the Small Business Administration of the United States government, the pamphlet is aimed at both prospective and active manufacturers' representatives. The definition's richness of detail and familiar example make it interesting, vivid, and easy to understand.

1   The manufacturers' sales representative, also known as a manufacturers' agent, is a professional salesman who is in business for himself. He sells compatible (not competing) products or, sometimes, services for one or more companies in a specified geographical territory. He and the companies he represents agree on both the territory and the markets, for example, retailers, wholesalers, and jobbers, to which he sells. In managing his own selling agency, the representative sells on a commission which is paid by his principals—the companies whose products he handles.

2   Products sold by representatives range from drugs to textiles. Among them are housewares, hardware, paint, chemicals, food processing equipment, electronic and electric components and products. Screws, bolts, nuts, steel, packaging—virtually all kinds of products may be sold through representatives. Services sold by representatives also span a wide spectrum. Examples are painting and plating services, machinery rebuilding and reconditioning, cleaning services, and business services.

3   The representative's arrangement with his principals is set forth in a written contract with each company that he represents. He cannot obligate his principals in any way without prior approval. Thus, he is not an "agent" in the strict sense of the word.

A Typical Representative

4   A typical manufacturers' representative is seen in the following example. William Trundle (name disguised) sells screw machine parts made by one company, perforated metals produced by a second company, gears manufac-

tured by a third company, and threaded fasteners fabricated by yet a fourth. His customers include design engineers and purchasing agents of manufacturing firms who produce consumer and industrial products.

5   His sales territory is Eastern Pennsylvania, Southern New Jersey, and part of Maryland. None of his product lines compete with another, and potentially he can sell quantities of each line to every customer he visits. He pays his own expenses and manages his time.

6   The representative does not take title to the goods he sells nor does he directly bear responsibility for providing credit to his customers. When the representative makes a sale, the product is shipped by his principal direct to the customer. The customer pays the principal for the merchandise, and the representative receives his commission after shipment and invoicing.

Representative's Qualifications

7   Many successful representatives have come from the sales management field, from purchasing, from engineering, and from servicing. Increasingly, recent graduates of schools and colleges are entering the field.

There are four basic considerations in becoming a sales representative: (1) desire to attain business independence, (2) ability to sell and manage one's own business, (3) knowledge of either a particular product line, specific markets, or geographical territory, preferably all three, and (4) strength, both financial and emotional, to sustain a sales agency through its first year or until success is achieved.

8   He may draw upon outside help for advice and guidance, but his own capabilities are the cornerstone of his business. He must become a specialist in what he sells. He must be able to gain the confidence and respect of his customers and his principals.

Advantages and Disadvantages

9   As in any small business, there are advantages and disadvantages to being a manufacturers' sales representative. The advantages for the representative include: independence, profit directly proportional to results achieved, opportunity for high income, ability to select the kind of people he associates with, and pride in building his own sales agency.

10   The disadvantages include: a lack of security equal to other businesses (earnings based on business activity rather than a salary), loneliness in the business community, and long working hours. Another disadvantage is the difficulty in building a business that can be sold in the event of an early death or upon the sales representative's retirement.

Questions for Analysis and Discussion

1. The purpose of this selection is to introduce readers to the position of the manufacturers' sales representative. What elements of the defini-

tion show that Warren assumes his readers have little previous knowledge about this field?

2. How does Warren make his definition concrete and down-to-earth? Why are these qualities particularly important, given the audience and purpose of this pamphlet?

3. The introduction to this chapter lists a number of devices that can be used to extend a definition (pp. 9–10). Analyze this definition to discover how many of these devices are used. Discuss how each clarifies the picture of the manufacturers' sales representative.

4. Technical writing is often persuasive as well as informative in its intent. Identify both the persuasive and informative aspects of this work, and discuss what effects they might have on the reader.

Application

Select a term in your field. Write a formal definition, then extend that definition in as many ways as you can, using the techniques listed in the introduction to this chapter (pp. 9–10). In the margin of your paper, write the technical term (as explained in the introduction) for each device every time it is used. ■

Crude Oil

The following definition is taken from *Refinery Siting Workbook,* prepared by the Mittelhauser Corporation for the Department of Energy. The workbook summarizes regulations and procedures that govern refineries. It was written to help governmental agencies and refinery officials reach timely decisions about refinery sitings, modifications, or expansions proposed by the industry.

This selection relies heavily on classification. Classification is often a useful technique in the early stages of writing because it can suggest a variety of subtopics that a writer may wish or need to cover. In fact, many writers begin by classifying a topic into its secondary subtopics and then use the resulting formal or informal outline to organize the material.

1   Crude oil is a mixture of compounds composed of carbon and hydrogen called hydrocarbons, and various amounts of sulfur, nitrogen, oxygen, trace metals, and water. The physical characteristics of a given crude oil can range from an almost clear liquid, similar to gasoline, to a pitch that is so viscous it must be heated to be pumped. Crude oil from geographically related oil fields tends to have similar compositions and properties.

2   Crude oils are typically designated as being sweet or sour and light or heavy. In addition, they are classified as being paraffin, intermediate, or naphthene based. For the purposes of this workbook, sweet and sour crude oils are defined according to the total sulfur content of the crude oil. While the refining industry has various definitions of sweet and sour crude oil, sweet crude oil refers to crude oil containing 0.5% or less of total sulfur (by weight). Sour crude oil contains more than 0.5% total sulfur. Likewise in this workbook, light crude oil is defined as having an API gravity of greater than 25. Likewise, heavy crude oil has an API gravity of 25 or less. The crude oil base depends upon the predominant type of hydrocarbon present. For example, extremely light crude oil (API gravity greater than 40) is generally paraffin based. Crude oils with API gravity between 40 and 25 are intermediate based, and crude oils with API gravity less than 25 are generally naphthene based. The vast majority of crude oils produced are of intermediate or naphthene bases.

3   The hydrocarbons present in crude oils may be separated by distillation into various fractions. While a fraction may contain many different hydrocarbon compounds, each fraction has the distinction of boiling within a specific temperature range. Lighter crude oils will produce a large proportion of frac-

tions that have lower boiling points (light fractions) and heavier crude oils will have a large proportion of fractions that boil at higher temperatures (heavy fractions). It is these fractions that are further processed into the final products. By varying process conditions, and unit processes, many different fractions can be produced, depending upon the final product requirements.

4   Of the non-hydrocarbon compounds present in crude oil, sulfur is the material of principal concern. Sulfur may be present as dissolved, free sulfur, hydrogen sulfide, or as organic sulfur compounds such as mercaptans. Generally, heavier crude oils will have higher sulfur concentrations, but sweet, heavy crude oils do exist. Other sulfur compounds, like hydrogen sulfide, may also be formed during the various refining processes. Sulfur compounds can cause severe corrosion to process units and can be a major source of air pollution.

5   The nitrogen and oxygen can also cause corrosion of process units and the nitrogen can also form ammonia which can cause violation of the wastewater discharge regulations. The trace metals, such as arsenic, nickel, and vanadium, may be present in such concentrations that they are poisons to certain process catalysts. The properties of the crude oil are all considered when deciding upon the overall refinery design basis.

Questions for Analysis and Discussion

1. This brief definition of *crude oil*, its types, and its characteristics, is designed for persons concerned with the location of oil refining facilities. What features of the definition indicate that the topic has been approached with this purpose in mind?

2. Outline the definition. How logical is its pattern of organization? How easily can the reader follow that organization? Can you think of anything the writer might have added to make the principle of organization clearer?

3. Even this highly technical definition employs a number of effective rhetorical devices, such as metaphor, analogy, and varied sentence structure. Point out some of these devices and discuss their effect in a definition such as this.

Application

How many ways is classification used in this example to help define *crude oil?* Select an important term in your field and classify it according to as many criteria as possible. Incorporate your system of classification into an informal outline.

J. Michael Polich and
Bruce R. Orvis
Rand Corporation

Alcohol Problems

The *Rand Reports* document the Rand Corporation's major research findings and results. This discussion of alcoholism comes from such a formal and comprehensive research report entitled *Alcohol Problems: Patterns and Prevalence in the U. S. Air Force* and was prepared for the Air Force. As an introduction to their study, the authors provide a literature review—that is, an evaluative summary of past studies—concerning the different meanings of *alcoholism.* In its treatment of past research, this excerpt demonstrates how sources are incorporated into a text. The selection also underscores how the same phenomenon—here, alcoholism—can have a variety of meanings, depending on the context in which it is used. In so doing, the discussion shows how the use of *differentiae,* or distinguishing characteristics—in this case, dependence, damage, or consumption—can lead to divergent definitions. Thus, when formulating their own definition, Polich and Orvis are careful to designate the "proper criteria for severe alcohol problems" that will be relevant for their study.

1   The scientific literature on alcohol problems contains a vast array of definitions, measures, and methods for counting and classifying the behaviors that fit the term. Virtually all definitions reflect a judgment that alcohol problems are "problems" because they are injurious; that is, because they damage or disrupt the individual or his associates. Apart from this, studies diverge on the question of what should be included under one heading.

2   A striking range of alcohol-related behaviors could be classified as injurious. At one extreme are behaviors that are clearly dangerous to health. In this class are such patterns as consuming alcohol at an extremely high rate over a long period, leading directly to liver disease. At the other extreme are behaviors that may have no determinate damaging consequences for the individual but are viewed as socially unacceptable by one's spouse, friends, or employers. Compounding this confusion is the fact that many kinds of alcohol-related behavior are viewed as injurious only in certain situations; for example, drinking any alcohol on the job is usually proscribed in the United States but not in many other countries.

3   These circumstances have led to a proliferation of definitions and doctrines on the subject of alcohol. No single interpretation is universally accepted. Indeed, there is intense debate over such basic definitional questions as whether the most serious manifestations of alcohol disorders should be treated as "diseases" (Keller, 1976; Robinson, 1972; Room, 1972). Some order may be brought into this confusion by distinguishing two main traditions in the prevalence literature: a tradition of studies of *clinical alcoholism,* and a sequence of more recent studies of *problem drinking.*

Clinical Alcoholism

4   ***Central Components of Alcoholism Definitions***   Alcoholism, treated as a clinically observed syndrome of problems associated with alcohol, has a multiplicity of definitions arising from its long history as a focus of humanitarian, social, and medical concern. In earlier periods, many accounts simply treated "drunkenness" as a moral problem without differentiating individual instances of intoxication from more chronic or severe manifestations of a continuing disorder (Keller, 1976). In this simple model, amount of alcohol consumption and intoxication were the subjects of interest. By the early 20th century, however, medically and psychologically oriented researchers were beginning to construct a series of different criteria for an "alcohol illness." These observers were impressed with the apparent compulsion of certain heavy drinkers to continue excessive consumption of alcohol despite serious consequences and even despite their expressed intention and desire to stop. Because of an apparent overwhelming need, or a "morbid insatiable craving" for alcohol (Paredes, 1976), such drinkers could be described as possessed of a psychiatric condition, variously described as "dipsomania," alcoholism, or (slightly later) alcohol addiction. Thus, the notion of *inability to control drinking* became a central part of the conception of alcoholism very early.

5   The other central component of most historical conceptions of alcoholism is the notion of *damage caused by alcohol,* especially physical damage leading to observable symptoms of functional impairment. Although clergymen, social workers, and psychologists have emphasized the social and behavioral damage excessive alcohol use can cause, medical doctors have exercised greater influence on definitional matters. Accordingly, many definitions of alcoholism concentrate on the physical sequelae of heavy alcohol consumption, such as liver disease and central nervous system disturbances. That this conception is still as powerful as ever may be seen from the "definition of alcoholism" recently offered in the *Annals of Internal Medicine,* which states succinctly that alcoholism "is characterized by tolerance and physical dependency or pathological organic changes, or both—all the direct or indirect consequences of the alcohol ingested" (National Council on Alcoholism, 1976). Obvious in this formulation is the primary role played by physical *consequences* of alcohol consumption. This emphasis on alcoholism's effects rather than on the behavior that constitutes alcoholism is frequently found in medically oriented research.

6   **Addiction and Loss of Control**   Elements of these primary components were interwoven into the theoretical formulation of the most influential author in the field, E. M. Jellinek. In proposing the "disease concept of alcoholism," Jellinek (1960) suggested that alcoholism might be treated as a disease with a biological basis in certain physiological alterations. Jellinek described these alterations as increased tolerance to the drug, adaptive cell metabolism, and the appearance of withdrawal symptoms when the drug is no longer taken. The crucial signal of the disorder was "loss of control"—the alcoholic's inability to moderate or stop drinking despite the most sincere desire to do so. In this view, the ingestion of any alcohol begins a reaction in which a physical demand or need for alcohol is felt ever more strongly.

7   These ideas, supplemented with informal data from Alcoholics Anonymous members, were the basis of Jellinek's elaborate theory, a notable aspect of which was the postulation of phases of alcoholism development. These were thought to begin with alcoholic blackouts and preoccupation with alcohol, to lead through the development of loss of control, and to end in a final stage characterized by physical deterioration, unemployment, loss of family and friends, and other adverse consequences associated with clinical alcoholism. Even though Jellinek proposed this theory as a working hypothesis, it immediately became the preeminent model for definition and diagnosis of alcohol problems.

8   **Conceptions of Dependence**   The notion of "alcohol dependence" was introduced partly as a euphemism for "addiction," as the criterion for the most severe alcohol syndrome. It achieved a new status when the World Health Organization (1952) adopted it in its definition of alcoholism. Partly at Jellinek's instigation, the WHO at that time declared:

> *Alcoholics are those excessive drinkers whose dependence upon alcohol has attained such a degree that it shows a noticeable mental disturbance or an interference with their bodily and mental health, their interpersonal relations, and their smooth social and economic functioning, or who show the prodromal signs of such development.*

In this view, then, alcoholics are a special subset of "excessive drinkers" (any drinkers whose drinking deviates from the community norms in quantity, frequency, or circumstance). The peculiar feature of alcoholics, setting them off from other excessive drinkers, is their dependence on alcohol. Although it has never been entirely clear, in this usage the term "alcoholics" would seem to include only those dependent people who actually experience adverse effects ("interference with bodily or mental health," etc.). Thus, this influential definition appears to require three elements for alcoholism: deviant drinking, dependence, and adverse effects of drinking.

9   The most nebulous concept in this formulation is the notion of dependence. Deviant drinking and adverse effects can be observed, however relative they may be in different social environments; but dependence lacks a clear measure. If the definition of dependence is not to rest almost exclusively on the subject's self-report that he desires alcohol, such a measure is essential.

In recent years an increasingly popular measure has been that of *physical dependence,* characterized by the appearance of a withdrawal syndrome when alcohol use is reduced or terminated. The symptoms of alcohol withdrawal are gross tremor, hallucinations, seizures, and delirium tremens in acute cases; milder cases have many other less specific symptoms (e.g., nervousness and sleeplessness). Recent physiological research suggests that the attainment of high blood alcohol concentration is a crucial aspect in the process of developing physical dependence and withdrawal symptoms (Gross, 1977). Thus, the concept of physical dependence on alcohol is a useful criterion for alcoholism, and the occurrence of withdrawal symptoms is a serviceable indicator in empirical studies.

10   Dependence is closely linked to another physiological phenomenon, *tolerance,* which refers to the body's ability to function in an outwardly normal manner even in the presence of high concentrations of ethanol (absolute alcohol). The most widely distributed diagnostic scheme, that proposed by the National Council on Alcoholism (1972), treats as a "classical" and "definite" indication of alcoholism *either* the appearance of withdrawal symptoms *or* the evidence of tolerance. According to this scheme, tolerance is indicated by a blood alcohol concentration of .15 without obvious intoxication. The judgment of what constitutes intoxication, however, is so subjective that this criterion has not as yet received much use in the empirical literature.

11   *Other Indications of Alcoholism*   We have discussed only the indicators that are most important for conceptions of the nature of alcoholism and alcohol dependence. Other indicators are frequently used in practice because of their status as strong correlates of alcoholism. Most prominent among these are various disease complications linked to alcohol consumption (e.g., alcoholic hepatitis or cirrhosis) and "blackouts" (memory lapse about events occurring during drinking the day or night before). Filstead et al. (1976) reported the ratings of such indicators, in terms of their usefulness for diagnosis, given by a sample of 362 physicians belonging to a U.S. medical society concerned with alcoholism. Over two-thirds of the group endorsed both the disease complications and the occurrence of blackouts as definite indicators of alcoholism. In the same sample, a similar proportion recognized all of the other criteria of dependence mentioned above (tremors, tolerance, subjective loss of control, etc.).

12   *Definitions Based on Consumption*   In all of these recent conceptions the actual amount of alcohol consumed by the individual plays a fairly minor role. The narrow context of the Jellinek theory makes this apparent anomaly comprehensible, because Jellinek was at pains to distinguish addicted drinkers ("real" alcoholics) from other excessive drinkers. Not being addicted, other excessive drinkers could be controlled through normal social mechanisms of education, law enforcement, etc.; but the addicted drinker, by definition, could not control his consumption and hence was unreachable by traditional sanctions. In this black-and-white world, amount of consumption

made little difference. The addicted alcoholic taking just one drink was in much more danger than the "chronic habitual excessive drinker" taking ten drinks.

13   Partly because of the divergence between this conception and the empirical evidence on alcohol consumption patterns, a school of thought has emerged recently that seeks to reemphasize the importance of amount of consumption (Schmidt, 1976). Loosely known as the "single-distribution" model, the theory advanced by this group derives its force from the strong aggregate correlations between cirrhosis mortality rates and mean per capita alcohol consumption in many populations. Numerous studies treating both cross-sectional and longitudinal international comparisons have shown that the level of mortality due to cirrhosis in a population is strongly related to the mean per capita alcohol consumption in the same population. . . .

14   The "single-distribution" theory has been used primarily to argue that the rate of "chronic excessive consumption" can be reduced by controlling the mean per capita consumption rate in the whole population. However, its implications for the definition of alcohol problems are also very important. It implies that at certain levels of consumption, serious adverse consequences become quite likely. Whether people showing those levels should be termed "alcoholics" is a semantic question. Nevertheless, it is clearly important to distinguish such people because of the adverse effects they are likely to experience.   ∎

15   *Dependence, Adverse Effects, and High Consumption*   Given these diverse definitions and viewpoints, what can we conclude about the proper criteria for severe alcohol problems? First, we should emphasize that there are at least three important and conceptually independent factors in drinking behavior that have historically been confused or combined:

1. *Alcohol dependence,* recognized primarily by physical dependence (withdrawal symptoms and/or tolerance) and loss of control;
2. The other *adverse effects* of heavy alcohol consumption, such as physical diseases (cirrhosis, hepatitis, cerebellar degeneration, etc.) and psychological and social impairments (unemployment, loss of family and friends, trouble with police, etc.); and
3. *Alcohol consumption,* the total quantity of ethanol consumed per day.

We have tried to separate these in the discussion. The original Jellinek formulation emphasizes dependence, although adverse effects are also mentioned. The 1952 WHO definition requires all three—dependence, adverse effects, and heavy (or at least deviant) consumption. The NCA diagnostic criteria accept any manifestation of either dependence or serious adverse effects, especially medical effects. Finally, the single-distribution school emphasizes heavy consumption, but perhaps admits serious medical conditions as well.

16   Several recent writers have recognized the conceptual confusion of this area and have pleaded for a clear distinction between the *condition* of alcohol

dependence and the *harm* caused by either heavy consumption or dependence. Davies (1976) suggests that alcoholism be defined as alcohol use that results in either dependence or substantial harm. Edwards (1976) also argues for defining dependence separately from the harmful consequences of alcohol use. Both commentaries avoid the term "alcoholism," which has been used in so many diverse ways that it has taken on excess meaning. Such a view finds increasing acceptance. The most recent evidence of the trend in this direction is the report of a new expert committee on definitions for the World Health Organization (Edwards et al., 1977). The committee explicitly avoided a definition of the term "alcoholism," preferring to talk instead about manifold "alcohol-related disabilities."

17   In this committee's view, the central disability related to alcohol is that of dependence—a chronic reliance on alcohol characterized by alterations of behavior away from normal patterns in consumption, subjective state, and physical state. Apart from dependence, the committee saw no particular commonality among all of the other disabilities that alcohol can cause. In particular, it emphasized that empirical evidence does not warrant an assumption that nondependent people with some "drinking problems" will necessarily progress into full-blown dependence (Edwards et al., 1977).

18   Our view is very much in sympathy with the conceptual distinctions advanced by this most recent WHO committee. The notions of alcohol dependence and the harmful consequences of alcohol are conceptually independent dimensions and should be addressed separately. Serious consequences can exist without any apparent dependence, and vice versa (Davies, 1976). This fact is obscured if a group of "alcoholics" is defined to be coterminous with one of the categories (or if the definition contains only their intersection). Scientific research is better served by an examination of the various phenomena and the interrelations among them.

19   All the definitions discussed above were developed through clinical experience, emphasizing alcohol dependence and its associated disabilities. Because these are the most serious manifestations of alcohol problems, this emphasis may be justified from a clinical point of view. In a study of prevalence rates, however, the sizes of various groups are of primary importance, and the alcohol-dependent group is very small. Many more people in any population are affected by alcohol problems of the nondependent kind than are affected by dependence. An assessment of the extent of alcohol problems in any population must therefore examine the other ways in which people get into trouble because of alcohol use.

——— *Polich and Orvis next review literature concerned with the other end of the spectrum of alcohol abuse, "problem drinking." They find that, unlike "clinical alcoholism," problem drinking "fail[s] to hold together into any coherent 'syndrome.'"*

The report then describes the few studies of alcohol-related problems conducted among other members of the military. On the basis of this review, Polich and Orvis conclude that "assessment of alcoholism rates [among Air Force personnel] will require development of new methodologies."—eds. ———

Conceptual Approach of This Study

20 The approach of this study has much in common with the literature just cited, but it also has a somewhat different emphasis. Our objective is not to isolate a particular clinical syndrome, nor is it to describe any and all types of alcohol problems that Air Force personnel may have encountered. Rather, we intend to isolate and identify groups of people who are *seriously affected by alcohol to the extent that official intervention may be appropriate.* The effect may be one that harms or seriously threatens to harm the individual, his immediate family, or the Air Force. Alcohol dependence, damage done to the person's health, accidents he may have, family problems, lowered productivity, or the necessity for increased law enforcement are instances of alcohol problems that fall under our purview, because all imply possible intervention.

21 Unlike many previous studies, this one attempts explicitly to distinguish two basic types of alcohol problems as follows:

- *Alcohol Dependence.* A chronic behavioral pattern indicating that the individual consumes high amounts of alcohol and relies on alcohol in everyday functioning.
- *Adverse Effects of Alcohol.* Any type of serious consequence of drinking not reflected under alcohol dependence if it results in concrete and serious damage or disruption to the individual's life or to the Air Force.

In general, we expect that people identified as "alcohol dependent" will show much higher levels of alcohol consumption, physical damage, work impairment, and chronicity of the condition. In contrast, those identified as having "adverse effects" should show lower rates of these problems and more intermittence in the condition. Alcohol-dependent people will be those for whom intensive treatment may be most appropriate.

22 In developing criteria to distinguish these two groups from the remainder of the Air Force population, we are guided by two considerations. First, we propose to use criteria that are concrete and minimally dependent on individual variations in attitudes and values. We do not, for example, wish to define as an "adverse effect" a behavior that does not cause serious trouble for the individual even though his spouse or his friends may object mildly to his drinking (or to any drinking). Our criteria are to be confined to *prima facie* evidence of damage (e.g., alcohol consumption at levels high enough to cause liver damage or that clearly interferes with work); or those that are so repugnant to the community that outsiders take drastic action (e.g., neighbors call police or spouses leave the subject because of drinking). Second, we propose to use policy-relevant criteria. At a minimum, the categories we distinguish should imply different intervention strategies to correct the problem. The criteria should not rest exclusively upon subjective or introspective judgments by the individual. Finally, the criteria should be generally useful in diagnosis, identification, or classification.

Bibliography

Davies, David L., "Definitional Issues in Alcoholism," in Tartar and Sugarman (eds.) (1976), pp. 53–73.

Edwards, Griffith, "The Alcohol Dependence Syndrome: Usefulness of an Idea," in Edwards and Grant (eds.) (1976), pp. 135–156.

Edwards, Griffith, et al., *Alcohol-Related Disabilities,* Offset Publication Number 32, World Health Organization, Geneva, 1977.

Filstead, William J., Marshall J. Goby, and Nelson J. Bradley, "Critical Elements in the Diagnosis of Alcoholism: A National Survey of Physicians," *Journal of the American Medical Association* 236: 2767–2769, 1976.

Jellinek, E. M., *The Disease Concept of Alcoholism,* Hillhouse Press, New Brunswick, N.J., 1960.

Keller, Mark, "The Disease Concept of Alcoholism Revisited," *Journal of Studies on Alcohol* 37: 1694–1717, 1976.

National Council on Alcoholism, "Criteria for the Diagnosis of Alcoholism," *Annals of Internal Medicine* 77: 249–258, 1972.

———, "Definition of Alcoholism," *Annals of Internal Medicine* 85: 764, 1976.

Paredes, Alfonso, "The History of the Concept of Alcoholism," in Tartar and Sugarman (eds.) (1976), pp. 9–52.

Robinson, David, "The Alcohologist's Addiction: Some Implications of Losing Control over the Disease Concept of Alcoholism," *Quarterly Journal of Studies on Alcohol* 33: 1028–1042, 1972.

Room, Robin, "Comment on 'The Alcohologist's Addiction,'" *Quarterly Journal of Studies on Alcohol* 33: 1049–1059, 1972.

Schmidt, Wolfgang, "Cirrhosis and Alcohol Consumption: An Epidemiological Perspective," in Edwards and Grant (1976), pp. 15–47.

World Health Organization, *Report of the Second Session of the Alcoholism Subcommittee, Expert Committee on Mental Health,* Technical Report Series Number 48, World Health Organization, Geneva, 1952.

Questions for Analysis and Discussion

1. Like "Terrorists and Terrorism" (pp. 20–21), this article deals almost as much with the problems of defining the term as with the definition of the term itself. In what ways does this discussion of alcohol problems demonstrate that using different criteria may lead to different definitions of an apparently singular subject?

2. Given their conceptual approach to this study, why is it necessary for Polich and Orvis to describe past studies in such detail?

3. Notice that the authors use the first-person plural *we* throughout their discussion. Where do they use it, and what effect does it have?

4. Words have *denotations*—their conventional meanings—and *connotations*—associations and suggestions. How do the different terms designating

forms of alcohol abuse evoke different pictures in the reader's mind? For example, what are the connotative differences between *alcohol addiction* and *problem drinking?* Why has *addiction* been replaced by *dependence?* Why do Davies (1976) and Edwards (1976) claim that the term *alcoholism* now has "excess meaning" (p. 37)?

Application

Many types of technical reports begin with an *informative abstract* or an *indicative abstract* (also called *descriptive abstract*) that either summarizes or indicates the report's content. (See pp. 166–67 for a fuller explanation of these abstracts; the article "Dwarf Sumac as Winter Bird Food," pp. 201–204, includes an example of an informative abstract.) Write an informative abstract and an indicative abstract for this report on alcohol problems and point out the differences between the two.

2

DESCRIPTION

The purpose of much technical writing is to provide the reader with accurate information about the physical properties of both natural and man-made phenomena, as well as information about more abstract phenomena, such as concepts, plans, and programs. While *definition* establishes the meaning of words, *description* communicates the physical and quantitative characteristics of objects, devices, mechanisms, or processes. In practice, of course, definition and description often overlap, since what something is may be closely tied to how it appears or operates. For this reason, description frequently forms an important part of extended definitions, just as definition is usually included in technical descriptions.

In technical writing, the aim of description is to present as objective, complete, and clear an image as possible. Unlike subjective writing, such as personal essays and certain types of advertising, which may appeal to the reader's emotions, technical description does not deal in impressions or feelings. Rather, it emphasizes concrete characteristics that can be quantified, measured, or analyzed. Imprecise or judgmental terms, such as *long, pretty,* or *cheap,* are replaced by specific terms or exact measurements, such as *two meters, hexagonal,* or *$1.59.* Although figurative language, particularly analogy, is sometimes included in technical description, its function is to help the reader visualize the phenomenon more clearly rather than to evoke feelings or attitudes.

Technical description encompasses both stationary phenomena—such as objects or static mechanisms—and active phenomena—such as processes or dynamic mechanisms. The primary difference between these types of description lies in the perspective from which each phenomenon is viewed: A stationary object or mechanism is described as it exists in space or at a particular moment, a process or dynamic mechanism as it changes and develops through time.

41

Generally speaking, descriptions of static, or unchanging, phenomena begin with a definition, then proceed to catalogue the parts of the object, explain how each part relates to other parts, and show how all parts combine to make up a unified whole. Descriptions of this sort often focus on physical attributes: size, shape, weight, density, color, composition—even model number and manufacturer's name can be important elements. The discussion in "High Clouds" (pp. 48–51), for example, provides precise measurements of each cloud's possible thickness, location, temperature, and water content, together with a detailed representation of how the cloud might appear to an observer. When the description is of a mechanism, like that of the pen (pp. 44–45), the writer will usually begin by explaining the mechanism's function and principle of operation—that is, the way it works. In every case, rational analysis is important: An object must be considered in a logical and systematic manner so that each part can be completely described and so that the reader can easily follow the progression of the description. Thus, the description of high clouds is divided into three sections, according to genus.

Descriptions of processes include much of this same information, often defining the process and explaining its function, theoretical background, or principle of operation, as well as the physical characteristics of any necessary mechanisms or materials. And these descriptions also aim for accuracy and precision. The major distinction between descriptions of static and dynamic phenomena is that a description of a process recounts a series of actions to explain how something happens, for instance, how wastewater is treated. The technique of recounting a series of related events—usually in chronological order—is called *narration;* this technique is used extensively in process descriptions. The writer will analyze the process in terms of cause and effect, before and after, and will divide it sequentially into major and minor stages, steps, or operations. The description of oil shale processing (pp. 66–68) illustrates this division into the separate stages of the production process and uses flow charts to reinforce the reader's understanding of how each stage relates to those preceding or following it.

The distinction between objects and processes is a useful one, but it does not always hold. For instance, examine the description of a data base (pp. 70–76), which is both an abstract concept and a method of storing and retrieving information. The writers describe the characteristics of a data base—what it is—as well as the processes by which it is created and accessed.

The importance of precision and accuracy in technical descriptions, whether of objects or processes, cannot be overemphasized. In some cases it is absolutely essential that not a single bolt be left unmeasured or unaccounted for—as in the case of specifications. *Specifications* are a specialized form of technical description that provides a verbal blueprint of anything from a small home appliance to a large-scale building project. The primary purpose of specifications is to define the standards to be met in the making of the object. Specifications are thus crucial documents in the design and manufacture of all kinds of products, as well as in engineering and construction. Indeed, in many instances, specifications form part of a legally binding contractual agreement (see the specifications for materials to be used for soil stabilization in the Beaver Creek development, pp. 54–55). When part of a contract, specifications spell out exactly and in extensive detail how work is to be done or how a product is to be made. Specifications for a large project, like the Beaver Creek village hall, may run to several hundred pages. In a more abbreviated form, specifications can be included in advertising, promotional, and product support literature. The specifications for the PHILIPS 2001 word processor (pp. 56–57) are an example of such use of this specialized form of technical description.

Like technical definition, description often forms part of larger works, such as proposals or reports. And as in all technical writing, the degree of detail and complexity found in a description is determined by the audience and purpose for which it is written. The reader must be supplied with the information that suits his or her particular expertise, background, and needs.

Pen

Many of the common objects we take for granted are as much a part of the technological universe as the most technically sophisticated processes and equipment. Often, we may know more about special technological inventions, particularly those we work with as professionals, than we know about the simple objects that surround us: toasters, zippers, light switches, and—as in the following selection—pens. This description of the pen is taken from *How Things Work,* a volume prepared by the editors of the *Encyclopaedia Britannica.* The aim of *How Things Work* is to explain a number of familiar machines and devices, and the principles behind them, to ordinary people. This selection also shows that the line between description and definition is not always clear.

1   The pen is an instrument for writing or drawing by transferring ink from the point of a stylus to paper. The earliest pens were brushes or quills, and the most recent are felt-tipped, but the two general types are fountain pens and ball-point pens.

2   Fountain pens are hollow plastic cylinders with writing points, or nibs, of various thicknesses at one end (Figure 1). The pen contains a flexible rubber or plastic saclike reservoir that holds the writing fluid, commonly an ink. The reservoir is filled by using an external lever or plunger, which acts in much the same manner as an eye dropper. The pressure exerted by the plunger will

Figure 1   Fountain Pen

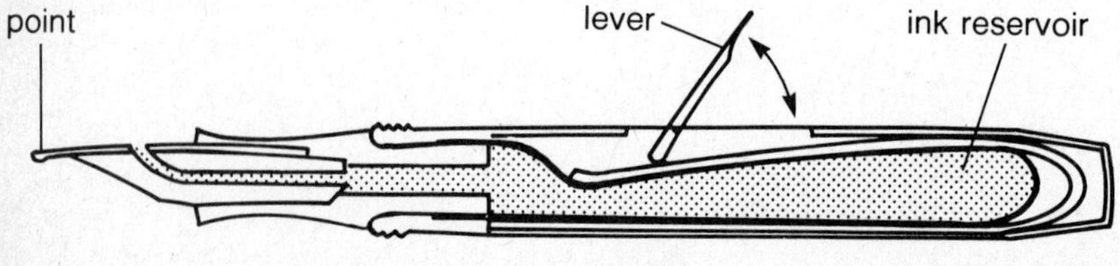

first evacuate the reservoir when the point of the pen is immersed in this fluid, and when released, the ink will be drawn into the sac. In operation, the ink from the reservoir passes downward to the nib during the writing process, and the time between refillings is determined by the capacity of the reservoir and the breadth of the desired line. Writing inks generally resist drying and thickening on the nib, but the nib is covered with a cap when not in use to avoid drying or evaporation of ink.

3   Ball-point pens operate in similar fashion, but, as the name implies, the writing point is a hard metal ball, commonly 1/25 inch in diameter, that rotates freely in a housing socket at the end of the pen (Figure 2). The ball is bathed in a constant supply of ink, derived from a slender tube that serves as an interior reservoir. The ink is held in the tube by capillary forces—produced by the adhesion of the ink to the wall of the tube—and when the ball turns it picks up ink, also by adhesion, and rolls it onto the paper. If the ball is smooth the ink may fail to adhere to it. For this reason, some balls consist of millions of tiny compressed particles, which produce an overall textured surface. The surface is in this way made up of many minute channels, which contain an ever-present supply of ink for instant writing. Many ball-point pens are designed to allow the point to retract into the body of the pen to keep it free of foreign matter. Several mechanisms are available for this function. Some pen points retract by pushing a button at the top of the pen housing; others retract by rotating the body of the pen. The retracting mechanism may be spring or ball-and-groove operated.

Figure 2   Ball-point Pen

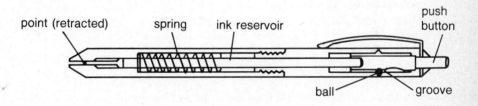

4   The inks used in ball-point pens contain a higher concentration of dye or pigment than do those intended for fountain pens, because the line laid down by a ball is much thinner than that from a writing nib. The time required to exhaust the reservoir of a ball-point pen is accordingly much greater. Of interest in this regard are tests conducted by the Parker Pen Company in 1976 to substantiate their advertising claims. A writing machine followed a figure eight for more than 25,000 feet using a medium point, a writing angle of 60°, and a rate of line generation of 262 inches per minute. Needless to say, the ink reservoir in such pens must be considered generous.

Questions for Analysis and Discussion

1. In what ways is definition important in this description of a mechanism? Identify the writers' uses of the formal definition and the various techniques they use to extend those definitions.

2. Outline the organization of this description. In what ways is it a logical way to arrange the material?

3. The editors of *How Things Work* describe their audience as "students . . . concerned citizens . . . the intellectually curious." On the basis of this excerpt, what knowledge do you think a reader needs to understand the book?

4. How do the writers keep the reader interested in what could be a dull subject? (After all, most people think they know how a pen works.)

5. Examine a pen to test the accuracy of this description. How would a description of your particular pen differ from this general explanation?

Application

Using similar techniques, describe another common object for a reader unfamiliar with that object. What are some of the problems of describing an object that is very familiar to you?

L. R. Koenig
and C. Schutz
Rand Corporation

High Clouds

In our everyday experience, we do not distinguish between cirrus, cirrostratus, and cirrocumulus clouds. Meteorologists, however, must be able to agree upon the characteristics of each type of cloud. Without such agreement, communication among scientists would be difficult and weather prediction impossible. This description of high clouds comes from the appendix to *Temperate-Zone Cyclonic-Storm Model,* a report prepared for the weather services of the U.S. Air Force by L. R. Koenig and C. Schutz of the Rand Corporation. The description helps Air Force meteorologists identify different types of clouds so that they are able to predict the possible formation of a cyclonic storm system.

Location: High clouds generally occur close to the tropopause. Consequently, their height is latitude dependent. On the average, their summits are 1.5 km below the tropopause; infrequently they occur in the lower stratosphere. They are often located in and ahead of frontal zones, regions of instability near the tropopause, jet streams, and in the aftermath of cumulonimbus activity. They may be initiated by orogenic waves.

Temperature: A representative cloud temperature is $-40°C$. The summit is commonly colder than $-40°C$ and the base is commonly colder than $-30°C$. Extensive streams of descending virga in some forms may cause lower portion of cloud to extend downward into warmer regions.

Cirrus (Ci)

Definition: Detached clouds in the form of white, delicate filaments or white or mostly white patches or narrow bands. These clouds have a fibrous (hair-like) appearance, or a silky sheen, or both.

Formation: Vertical motions at level of formation. Motions associated with billows or waves, frontal lifting, leading edges of anvils associated with cumulonimbus and jet streams. Wavelength of generating motions most frequently between 250 and 500 m. Very high frequency of occurrence over cyclonic regions and fronts. Very high probability of occurrence at warm and quasistationary fronts, somewhat less (but still high) probability along cold fronts. Occurrence in uniform air mass less likely.

Season: Any

Horizontal Extent: Highly variable, clouds associated with active fronts may be 2000 km along the leading edge of the high cloud shield associated with the front. Clouds associated with cumulonimbus activity and jet streams are less widespread and isolated patches are not uncommon.

Thickness: Thickness may vary from a few hundred meters to 7 km. A nominal value is 2 km. Frontal cirrus are generally thicker than air mass cirrus. Cloud thickness and tropopause height are correlated positively. Clouds associated with jet streams generally lie 1 to 2 km below the jet axis, rarely above. Clouds often composed of individual elements in a generation zone and have an extended virga region composed of cloud particles falling out of the generating zone. Particles generally fall into subsaturated air and slowly evaporate as they descend. However, they may renew growth by falling into layers subsaturated with respect to liquid but supersaturated to solid.

Homogeneity: Cloud water content is greatly variable over short distances both in the vertical and horizontal. This is due to varying particle fall velocity, their evaporation and growth and wind shear, as well as the patchy nature of the generating zone. Generating zones for cirrus uncinus ("mare's tail"), a common hooked-shaped form, are in the order of 1 to 10 km in diameter.

Hydrometeors:	Ice crystals—generally hexagonal columns, frequently with one end having pyramidal faces ("bullets"). Clusters frequent, composed of many (i.e., 10) crystals apparently jointed with pyramidal ends radiating from a common point.
	Particle size—100 to 200 μm in length, 20 to 50 μm thick, clusters proportionally larger (i.e., 200 to 500 μm in diameter).
	Liquid droplets—infrequent occurrence. There is speculation that liquid droplets are formed during the initial appearance of the cloud, but if this is true, they are rapidly converted to ice.
	Number density—10×10^3 per m³ is a representative value.
Water Content:	Nominal maximum value 0.3 g/m³. Average values are about 0.1 g/m³. Clouds are not homogeneous in either the vertical or horizontal dimension. Mixing, by turbulence and gravitational separation, will generally cause lesser water content at edges of cloud, great content in center, but small-scale fibrous nature of cloud indicates rapid, small-scale variation in water content. Cloud particles can fall a great distance in subsaturated air before completely subliming. Consequently, regions of extremely low, but finite, ice-water content are often characteristic of the cloud "base."
Turbulence:	Generally slight.

Cirrostratus (Cs)

Definition:	Transparent, whitish cloud veil of fibrous (hair-like) or smooth appearance, totally or partly covering the sky, and generally producing halo phenomena.
Formation:	Large-scale lifting at level of formation, creating a more-or-less uniform cloud mass. Motions may be caused by frontal lifting or convective currents associated with cumulonimbus. Cirrocumulus may undergo transformation to cirrostratus.
Season:	Clouds occur in any season but most cloudiness, being associated with traveling cyclones and frontal activity, occurs when these are most prevalent (cooler seasons in temperate latitudes).
Horizontal Extent:	Highly variable, clouds associated with fronts generally cover very large areas. Representative dimensions are 2000 km along the axis of the front and 1000 km normal to it (the area may be shared with other high cloud forms). Clouds associated with smaller source regions; such a cumulonimbus activity will have correspondingly lesser horizontal extent.
Thickness:	Clouds range from 100 m to about 7 km in thickness. A nominal value is 1.5 km. They tend to thicken nearer the source of vertical lifting. Multiple layers are not uncommon.
Homogeneity:	Cloud water content relatively uniform in comparison with other high clouds.
Hydrometeors:	Ice crystals—If caused by large-scale lifting, hexagonal columns, frequently with one end having pyramidal faces ("bullets"), will predominate. Clouds associated with cumulonimbus anvils will

contain ice particles formed by processes involving the freezing of
drops and will have irregular as well as regular shapes.

Particle size—If caused by large-scale lifting, particles will generally be
100 to 200 μm in length, 20 to 50 μm in thickness. Clouds
generated by cumulonimbus activity will contain a wider size
distribution (both smaller and larger size present). Gravitation
separation removes large material close to the source region and
evaporation continually removes small material.

Water Content: Nominal value 0.1 g/m³. Clouds generated by cumulonimbus activity will
have higher values close to the source; those associated with fronts or
air masses, lower values. Clouds are reasonably homogeneous, but
mixing and gravitational separation will generally cause lesser water
content at edges of cloud.

Turbulence: Generally slight.

Cirrocumulus (Cc)

Definition: Thin white patch, sheet, or layer of cloud without shading, composed of
very small elements in the form of grains, ripples, etc., merged or
separated, more or less regularly arranged; most of the elements have
an apparent width of less than one degree.

Formation: Formed either by regular cellular convection or at crests of low
amplitude, short wavelength eddies. Wavelength of eddy commonly
less than 250 m, rarely greater than 500 m. Waves may be orogenic or
caused by lifting, mainly along cold fronts, rarely warm fronts. They are
also found associated with jet streams, generally lying about 1 to 2 km
below the jet axis, rarely above it. Cloud elements that are composed of
small cumuliform tufts indicate instability at cloud level. Orogenic clouds
are more frequent under stable (nocturnal) than unstable (midday)
conditions.

Season: Any; extensive cloudiness mainly occurs during seasons in which frontal
activity is most frequent (colder season in temperate latitudes).

Thickness: Cloud may be several km thick.

Hydrometeors: Ice crystals—generally hexagonal columns, frequently with one end
having pyramidal faces ("bullets").

Particle size—maximum 100 to 200 μm in length, 20 to 50 μm thick.
Any liquid droplets formed during cloud generation are rapidly
converted to ice.

Water Content: Nominal maximum value 0.1 g/m³. Cloudlets are not homogeneous in
either vertical or horizontal dimension, but alignment of cloudlets in
regular patterns suggests more or less regular periodicity in cloud
properties such as water content and particle number density.

Turbulence: Generally light, but may become stronger in active, convective situations. Some observations of aircraft bumpiness have suggested vertical currents of 4 to 5 m sec^{-1} in cold-front cirrocumulus.

DISTINCTION AMONG GENERA

Cirrus: Cloud particles arranged in discontinuous units having a fibrous, hair-like appearance. Effects of vertical wind shear are often manifested by streamers descending from generating cells. Generating cell units are arranged with no regular pattern. Cloud water content and thickness are highly variable and irregular. Cloud base poorly defined because of virga.

Cirrocumulus: Cloud particles arranged in very small units (generating cells) having a patchy, herring-bone, or rippling appearance. Generally a regular patterned structure encompassing many units is discernible. Virga and effects of wind shear are not prominent. A cloud composed of many units may or may not cover the entire field of view from the ground. Cloud water content is highly variable, but has a more or less regular spatial pattern. Cloud top and base well defined. If generating cells become widely separated and display no regular pattern, transformation to cirrus is indicated. Cirrostratus transformation occurs if wave structure disappears.

Cirrostratus: Cloud particles more uniformly dispersed than in either cirrus or cirrocumulus. Therefore cloud properties are more uniform in both horizontal and vertical directions. Generally, a cirrostratus cloud covers a greater horizontal area than cirrus or cirrocumulus. Increasing uniformity of cirrus or cirrocumulus will cause transformation to cirrostratus.

Questions for Analysis and Discussion

1. This selection uses a good deal of figurative language to make the description more exact (for instance, the "silky sheen" of the cirrus cloud). Identify some of these uses and discuss their effects. What makes such language appropriate to a technical description? How does it help achieve an accurate representation of the phenomena described?

2. What elements of the format and organization of this description help the reader obtain specific information as easily and quickly as possible? How are format and organization a function of the intended use of the report?

3. This description is written for the expert. Can it be understood, at least in part, by the nonexpert? Can you identify high clouds on the basis of this description? What aspects of the description are helpful to you?

Application

Observe clouds or a cloudy sky. From your observations, write an impressionistic description of what you see. Compare your description with the language, tone, and organization of this selection. What do the differences reveal about technical description?

Specifications

Specifications are a special form of technical description. They provide a prescriptive list of the materials, dimensions, and acceptable methods of construction or assembly of many products. Frequently referred to as simply "specs," they can be written for anything from a small household appliance or piece of industrial equipment to a large-scale construction project—a highway, a hospital, a spacecraft.

Two sets of specifications follow. The first is excerpted from a lengthy set of specifications prescribing the materials and methods of construction for the Beaver Creek Village Hall, part of a Colorado resort and leisure development. It specifies the type of matting to be used to prevent erosion from landscaped areas surrounding the hall and the seed mix with which these areas are to be planted. This set of specifications was written by the architects to guarantee that the development is constructed exactly as designed. Because it forms part of a legally binding document, accuracy and precision are essential.

The second set of specifications describes a word processor and is supplied by Philips as part of their advertising literature.

Soil Stabilization:
Materials

2.01 MATERIALS

1   A. Jute Matting: Cloth of a uniform plain weave of undyed and unbleached single jute yarn, forty-eight (48) inches in width plus or minus one (1) inch and weighing an average of 1.2 pounds per linear yard of cloth with a tolerance of plus or minus five (5) percent, with approximately seventy-eight (78) warp ends per width of cloth and forty-one (41) weft ends per linear yard of cloth. The yarn shall be of a loosely twisted construction having an average twist of not less than 1.6 turns per inch and shall not vary in thickness by more than one-half (½) of its normal diameter. Jute mat to be Ludlow Soil Saver or equal. Install according to manufacturer's instructions.

2   B. Erosion Control Staples: Wire, 0.091 inches in diameter or greater, "U"-shaped with legs minimum six inches (6") in length and one inch (1") crown.

3   C. Erosion Control Tackifier: Ecology Control M Binder or approved equal.

4   D. Seeding Mix is available from:

Rocky Mountain Seed Company
1325 15th Street
Denver, Colorado (313) 623-6223

Seed mixture shall be at least 98% pure, noxious pest and weed free, with a minimum of 85% germination. All seed shall be re-cleaned Grade A 'new crop seed' delivered in the original containers, unopened and shall bear a guaranteed analysis and dealer's label. The seed mix shall consist of the following high altitude seed mix for elevations below 9500 ft.:

Type	Variety	Percent by Weight	Rate Pounds/Acre
Orchard	Potomac	13	6.5
Perennial rye		20	10.0
Winter wheat		20	10.0
Smooth brome	Manchar	27	13.5
Timothy	Common	13	6.5
Clover	White Dutch	7	3.5
Total		100	50.0

All leguminous seeds to be inoculated with appropriate bacteria.

PHILIPS 2001
Specifications

1   The PHILIPS 2001 is a modular, high performance, standalone word processing system. Individual components may be arranged as the operator chooses, thereby ensuring maximum comfort and ease of operation. All parts are cable-connected. Maximum cable length 10 feet.

OPERATING CONSOLE
Consists of electronic keyboard and CRT screen.

Keyboard
Integrated with CRT unit. Fully electronic. 47 alphanumeric keys.

CRT Unit
15 inch, diagonal display screen. 31 display lines; 3 control lines, including prompt/message, status, and format lines, plus 28 text lines. 80 characters per line. Scrolls 94 lines vertically, 250 characters horizontally. Flashing underscore cursor. Adjustable brightness. Green characters on black, no glare screen.

Dimensions
15 inches high × 18.9 inches wide × 23.6 inches deep. 38cm (h) × 48cm (w) × 60cm (d).

Weight
41.9 lbs (19Kg)

CONTROLLER
Houses entire processing unit, power supply, static power regulator and disk drives.

Central Processing Unit
Intel 8080A microprocessor with over 64,000 characters of memory or a Zylog Z80A with 128,000 characters of memory. Screen memory of 8,000 characters.

Disk Drives
Independent dual disk drive unit for disk storage and program loading. Disk drive may be used concurrently or independently of each other. Single Disk Drive unit available in the 64K version.

MICOM Diskettes
Pre-formatted, single sided, diskettes. 77 tracks. 32 sectors. Each diskette stores up to 300,000 characters or 127 pages of text.

Console Dimensions
24.6 inches high × 22.9 inches wide × 20.5 inches deep. 67cm (h) × 58cm (w) × 52cm (d).

Weight
75 lbs (34Kg)

PRINTER
Standard, Qume 15 inch, Bidirectional, Daisy-type printer. Prints 45 characters per second. Top-of-Form and End of Ribbon detect. Printing width:
13.2 inches
132 characters in 10 pitch.

158 characters in 12 pitch.
198 characters in 15 pitch.
96 character printwheels in a wide variety of fonts. Variable pitch, from 3 to 40 characters per inch. Variable line spacing—¼-line increments, up to 3-¾ line spacing. Fabric, multi-strike, and single-strike ribbon cartridges.

Print Features
Automatic justification and Bold printing. Background linear or non-linear merge.

Dimensions
7.9 inches high × 23.6 inches wide × 17.7 inches deep. 20cm (h) × 60cm (w) × 45cm (d).

Weight
50.7 lbs (23Kg)

POWER REQUIREMENTS
Voltage
115 A.C. 60Hz

Power
650 VA maximum

ENVIRONMENT CONDITIONS
Normal office environment.

SOFTWARE PROGRAMS
PHILIPS 2001's are fully compatible with all PHILIPS software programs including: Math Pak II, Sort, Record Processing, OCR, Keystroke Memory, Proportional Spacing, BASIC, Asynchronous, 2780 or 3780 Bi-synchronous and MICONET communications.

HARDWARE ACCESSORIES
Fully compatible with all PHILIPS hardware accessories including: Sheet Feeder, Continuous Forms Tractors, Wide Track and Dual Head Printers, OCR equipment and the Shared File System.

Questions for Analysis and Discussion

1. What qualities of the Beaver Creek Soil Stabilization specifications ensure that there will be no confusion about the materials and methods the authors want used?

2. The specifications for the PHILIPS 2001 appear, at first glance, simply to provide a technical description of the physical and performance characteristics of the word processor. However, they are excerpted from Philips' advertising material. Isolate those features that are clearly persuasive in nature—that is, designed to convince the customer to purchase the product.

3. Contrast the audience and purpose of the PHILIPS specifications with those of the Beaver Creek specifications. What features of each reflect the differences in audience and purpose?

4. Most common mechanical or electrical consumer items come with product support literature that includes a set of specifications. Look at

the material supplied with your automobile, stereo equipment, television, or small appliance. Are specifications supplied? How do those specifications compare with the sets included here?

Application

Select a common and fairly simple item with which you are familiar. Determine as thoroughly as you can the materials and dimensions of the object. Write a set of specifications for the object that is as accurate and precise as possible. Be certain to arrange the specifications logically, to make them easily comprehensible.

Celrobic Wastewater
Treatment System

These two descriptions of a wastewater treatment process are taken from two Celanese Corporation publications. The first comes from *Celanese World,* a magazine produced for employees and other persons interested in the corporation. The second comes from a brochure prepared for engineers who are considering implementing this wastewater treatment process. Both are adaptations of a highly technical paper, originally presented by the developer to fellow scientists. Thus, each of these descriptions gives a simplified account of the process, but each is directed to a different audience and serves a different purpose. These differences have determined the amount of knowledge the prospective reader is assumed to possess, as well as the level of detail and precision with which the process is described.

Environmental Protection: Celanese Technology Is a Step Ahead in Wastewater Treatment

Celrobic Wastewater Treatment System

1. A process developed by Celanese;
2. An economic alternative to conventional methods for treating industrial effluents;
3. A means of producing from organic wastes a methane-rich gas usable as a boiler fuel;
4. A process that produces significantly less solid by-product waste and requires less space;
5. A system applicable to most industries that produce wastewater containing organic chemicals.

1   Celanese Corporation always has been concerned with the quality of water it discharges from its plants back into the environment. This concern is evidenced by the development by Celanese Chemical Company, Inc. (CCC) of a new anaerobic wastewater treatment process.

2   This process represents the latest development in biological treatment for water quality control. And, Celanese has taken the lead with the first successful installation of an anaerobic process for treating industrial organic wastes.

3   CCC began evaluating various improved methods of treating plant wastewater in the early 1970s. At that time, CCC, like many other companies, had been using conventional aerobic systems to treat plant wastewater. In a conventional aerobic system, wastewater is mixed with air enabling oxygen-consuming bacteria to break down the organic waste. The aerobic method requires large quantities of electricity to mix the air and produces relatively large amounts of sludge that needs additional treatment.

4   Anaerobic processes are nature's most effective means of dealing with organic wastes. However, until recent development work, much of it done by Celanese, methods of applying these processes to industrial wastewater have been expensive and ineffective. These drawbacks were coupled with difficulties in process maintenance and control, making anaerobic treatment impractical for most industrial applications.

5   Anaerobic treatment also relies on bacteria to break down organic waste, but does so without oxygen. In the new Celanese *Celrobic* system, the biological action of anaerobic bacteria occurs in a closed reactor and reduces organic material in the treated water to a methane-rich gas.

6   The heart of the *Celrobic* process is a reactor filled with anaerobic bacteria through which water circulates. The bacteria adsorb and degrade the organic material in the wastewater. As the water flows out of the reactor, a portion of it is mixed and recycled back into the reactor with the incoming wastewater. In this way, incoming wastewater is diluted and prepared for treatment in the reactor.

7   Continuous monitoring of incoming wastewater, outgoing treated water and by-product methane gas enables system operators to maintain strict control over the *Celrobic* process. The process is equipped with on-line analyzers and proprietary control systems that automatically perform critical system control functions.

8   The *Celrobic* anaerobic system has several important advantages compared to the conventional aerobic systems. "With energy costs rising and fuel supplies dwindling," says Dr. Roger L. Van Duyne, *Celrobic* systems marketing manager, "the anaerobic process is ideal, because it not only utilizes much less electricity than conventional aerobic processes, but also produces methane gas which is collected and used directly as a valuable boiler fuel. It also requires much less space than aerobic systems and produces about one-tenth the amount of sludge."

9   The *Celrobic* process eliminates the disadvantages associated with early anaerobic systems in the treatment of industrial waste. The prototype process, which was installed at the Vernon, Texas, plant of Celanese Plastics & Specialties Company (CP&SC), removes the major portion of biodegradable organic contaminants in the plant's wastewater. The Vernon facility has been operating since 1977. CCC has two other *Celrobic* systems under construction at its Pampa and Bishop, Texas, plants. Both operations are due on stream in early 1981.

10   The *Celrobic* process used alone will treat high-strength wastewater and produce low contaminant levels. The outflow then can be sent directly to municipal aerobic systems for final treatment. By using aerobic treatment after the *Celrobic* system, organic waste is reduced to an extremely low level.

11   In the next decade, industry will more than double its investment in wastewater control, predicts Dr. Van Duyne. Because of the widespread need for economical water treatment systems, Celanese is licensing the *Celrobic* process for use by other companies. The process is applicable to practically any industry where high-strength organic wastes are generated. These include the petrochemical, food, tobacco, textile, brewing, distilling, and pulp and paper industries.

The Celanese Anaerobic Wastewater Treatment Process

Introduction

1 Anaerobic processes have been recognized for generations as nature's most effective manner of dealing with organic waste materials. Until recently, methods of applying these processes to the practice of wastewater treatment have involved very long detention times and high capital commitments. These problems coupled with difficulties in process maintenance and control have made anaerobic treatment impractical for most industrial applications.

2 After several years of research, Celanese Chemical Company, Inc. has developed the Celrobic™ process, a contact anaerobic treatment which overcomes disadvantages previously associated with anaerobic systems. This high rate process has been used in numerous laboratory and pilot plant studies to treat a wide variety of high strength chemical and natural product wastes.

3 The prototype Celrobic™ process, designed for 65% removal of a 36,000 lb/day COD (chemical oxygen demand) load, has been in continuous operation at the Celanese guar bean processing plant in Vernon, Texas since 1977. Celanese has two other plants currently under construction at chemical plants in Pampa and Bishop, Texas. The Bishop system is designed to remove 80% of a 75,000 lb/day COD load, while the Pampa system will remove 90% of a 117,000 lb/day COD load. These new units are scheduled to start up in early 1981.

The Celrobic™ Process

4 The Celrobic™ process is comprised of a packed anaerobic reactor through which water circulates in an up-flow manner. Bacteria develop on the contact medium, and the organic material is adsorbed and degraded by the entrapped organisms. Although liquid detention time is short, residence times of bacteria and organics are substantial. Overflow from the packed bed is continuously recycled to mix with influent to the reactor. In this way, incoming wastewater is diluted and prepared for the treatment area. The required recycle rate is dependent upon wastewater strength and its chemical composition. Biological action within the reactor reduces the organic material first to carboxylic acids and then to methane. Normally, methane gas collected from the system can be used directly as a valuable boiler fuel.

5 The amount of organic reduction obtained in the Celrobic™ process varies with waste composition and design of the reaction system. These reductions,

however, are usually not sufficient to provide complete wastewater treatment. Discharge from the anaerobic reactor may be sent to a municipal plant or other secondary aerobic treatment facility for final polishing.

6   Features of the Celrobic™ process are illustrated by the engineering model shown in Figure 1. Celanese holds patents on the process and additional U.S. Patents are pending.

Figure 1   Features of the Celrobic Process

Effluent Equalization Tank

Anaerobic Reactor

Sodium Carbonate Storage System   Sodium Carbonate Feed System   Influent Effluent Heat Exchangers   Feed Heater   Gas Compressors   Emergency Flare

Advantages of the Celrobic™ Process over Conventional Aerobic Treatment

7   **– Greatly Reduced Net Energy Requirements**
Electric power requirement for the Celrobic™ process is a fraction of that required for an aerobic plant. An aerobic plant typically requires about 1M KWH per ton of COD removed, while the Celrobic™ process requires only one-quarter of that energy or less, depending on the COD content of the water.

– Greatly Reduced Sludge Production
Sludge produced by the Celrobic™ process is approximately one-tenth of that produced by an aerobic process, thus minimizing the expensive problem of sludge disposal.

– **Methane Production**
The Celrobic™ process produces a methane-rich product gas at about 5.9 standard cubic feet of methane per pound of COD removed. When the gas is used as boiler fuel, an attractive operating cost credit is realized. Gas credits from treatment of high strength wastes are usually greater than total operating costs for the Celrobic™ system.

– **Treats High Strength Wastewater**
Anaerobic wastewater treatment is widely recognized as a superior alternative to aerobic treatment, particularly for high strength wastes (> 5000 mg/1 COD). Fewer constraints on food/biomass control is a key advantage. Also, the cost savings offered by the Celrobic™ process become greater as wastewater strength increases.

– **Synergy with Aerobic Polishing**
Anaerobic removal of biodegradable components from effluents that also contain refractories allows development of a specialized sludge in an aerobic polishing unit. As a result, Celrobic™ treatment followed by aerobic polishing often can produce final outfalls significantly lower in COD than can be accomplished by activated sludge without anaerobic pretreatment.

– **Greatly Reduced Nutrient Requirements**
Nutrient requirements are normally less than one-tenth that of an aerobic system. Many wastewaters contain sufficient nitrogen and phosphorous to eliminate the need for additional nutrients.

– **Lower Sensitivity Toward Heavy Metals**
Most heavy metals and related complexes are converted to insoluble sulfides in the Celrobic™ treatment process and are rendered harmless to the anaerobic population, or to downstream facilities.

– **Automatic Operation and Minimal Operator Attention**
To eliminate deactivation of the methane-forming bacteria due to organic, hydraulic, or toxic chemical overloads, the Celrobic™ process is equipped with on-line analyzers and proprietary automatic control systems. The system can be shut down for hours or even days and can be restarted with virtually no loss of activity.

– **Widely Varying pH Levels of Influent Tolerated Without Neutralization**
If caustic wastes are to be treated, they will be neutralized by the acid-forming action of the bacteria population. A carbonic acid-bicarbonate buffer formed in the Celrobic™ system minimizes the amount of caustic needed for neutralizing acidic wastewater. These factors lower the cost of chemicals and equipment for pH adjustment.

– **Lower Maintenance Costs**
Fewer pieces of rotating equipment (pumps versus aerators) allow savings in both maintenance and capital costs.

Applications for the Celrobic™ Process

8 The Celrobic™ process is applicable to practically any industry where organic wastes of 1000 mg/1 or higher are produced.

9 Examples are the following:

CHEMICALS	BREWING AND DISTILLING
PETROCHEMICALS	PULP AND PAPER
FOOD PROCESSING	TOBACCO PROCESSING
TEXTILES	

Questions for Analysis and Discussion

1. Compare these two descriptions of the *Celrobic* process in terms of their audiences. What aspects of content and language reveal how the information has been adapted for each group of readers?

2. Compare the readability of these two descriptions. Examine both for their use of well-formed sentences using subordination, parallelism, inverted word order, ellipsis, and other forms of complex sentence organization (these techniques are explained in the Foreword, pp. 1–7). Then calculate the average number of words per sentence (the average sentence length) for each description. What conclusions can you make concerning the contributions of sentence structure and length to the readability of each description?

3. Both of these publications attempt to do more than merely inform the reader. The first was written to demonstrate Celanese's concern for environmental quality. The second selection was written to sell the product. Point out the promotional features of each selection.

4. Select a topic discussed in the science, technology, or business sections of a popular news magazine such as *Time, Newsweek,* or *U.S. News and World Report.* Compare the features of such journalistic treatments of a particular topic with its treatment in a professional journal in the appropriate field. How is the journalistic report slanted to appeal to a general readership? Look in particular at content, graphics, word choice, and style.

Application

Select what you judge to be a highly technical article from a professional journal in your field. Rewrite it to suit the needs and interests of a general reader.

Oil Shale Processing

The Office of Technology Assessment is an agency of the U.S. Congress. The following description of oil shale processing comes from its report, *An Assessment of Oil Shale Technologies.* This two-volume study reviews methods of developing oil shale resources and evaluates their consequences. The report is designed to aid Congressional analysis of the potential benefits and hazards of oil shale production.

1   Converting shale in the ground to finished fuels and other products for consumer markets involves a series of processing steps. Their number and nature are determined by the desired mix of products and byproducts and by the generic approach that is followed in developing the resource. The alternative approaches are:

- TIS processes in which the shale is left underground, and is heated by injecting hot fluids;
- MIS processes in which a portion of the shale deposit is mined out, and the rest is broken with explosives to create a highly permeable zone through which hot fluids can be circulated; and
- AGR processes in which the shale is mined, crushed, and heated in vessels near the minesite.

Figure 1 is a flow sheet for the steps common to all three options. How the steps would be integrated in an AGR facility is shown in Figure 2. In the first step, the oil shale is mined and crushed for aboveground processing, or the deposit is fractured and rubbled for in situ processing. The main product is raw oil shale with a particle size appropriate for rapid heat transfer. Nahcolite ore can be one of the various byproducts from this step. Dust and contaminated water are among its wastes. In the retorting step, the raw oil shale is heated to pyrolysis temperatures (about 1,000°F (535° C)) to obtain crude shale oil. Other products are the spent shale residue, pyrolysis gases, carbon dioxide, contaminated water, and in some cases additional nahcolite and dawsonite ore. The crude shale oil may be sent to an upgrading section in which it is physically and chemically modified to improve its transportation properties, to remove nitrogen and sulfur, and to increase its hydrogen con-

Figure 1 Oil Shale Utilization

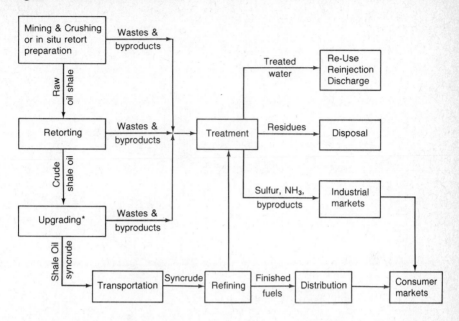

*May not be needed for some in situ methods.

tent. (Crude shale oils from some in situ processes may not need upgrading before transportation.) Contaminated air and water, and in some units refinery coke, are the wastes produced along with gases that contain sulfur and nitrogen compounds. Depending on the extent of the treatment, the upgraded product—shale oil syncrude—can be a high-quality refinery feedstock, comparable with the best grades of conventional crude. In the refining step, which may be conducted either onsite or offsite, hydrogen is added to convert the syncrude to finished fuels such as gasoline, diesel fuel, and jet fuel. The syncrude, or the crude shale oil, could also be used directly as boiler fuel. After refining, the fuels are distributed to consumer markets. Refining also produces waste gases and various contaminated condensates.

2   To protect the environment, contaminated water must be treated for reuse in the oil shale facility, for reinjection into the ground water aquifer source, or for discharge into surface streams. Contaminated air and process gases must be purified to meet Federal and State air pollution standards before they can be discharged to the atmosphere. The waste gases from retorting, upgrading, and refining are potential sources of ammonia (for fertilizer and other uses) and sulfur (for sulfuric acid and many other materials). These can be recovered during the treatment steps and sold to industrial processors. In some portions of the Green River formation, the solid residues from mining and retorting will contain nahcolite (which can be used to produce soda ash

Figure 2   The Components of an Underground and Aboveground Retorting Oil Shale Complex

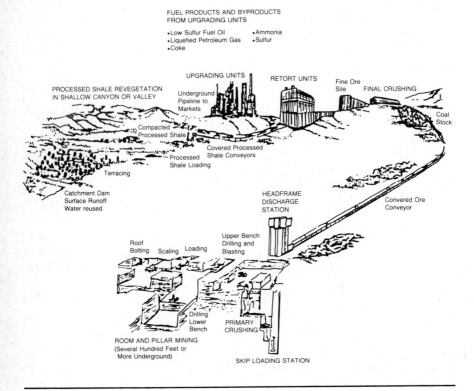

FUEL PRODUCTS AND BYPRODUCTS
FROM UPGRADING UNITS

- Low Sulfur Fuel Oil   • Ammonia
- Liquefied Petroleum Gas   • Sulfur
- Coke

and for stack gas scrubbing) and dawsonite (a source of aluminum metal). In any case, spent shale from aboveground processing must be moistened, compacted, and revegetated to prevent erosion and leaching. Retorted shale in situ retorts, and spent shale from surface operations that is backfilled into underground openings, must be protected from leaching by ground water.

Questions for Analysis and Discussion

1. This description contains highly technical information, yet it is addressed to readers who may not have specialized technical knowledge of the field. Has the writer adequately adapted the material to anticipate the problems an intelligent but technically uninformed reader

might have? Identify particular places where you feel the writer has or has not considered the general reader's background and needs.

2. This discussion of the process of converting shale to finished fuel describes a chronological sequence of actions. How does the writer help the reader follow the order of the steps? Point to specific words and phrases that indicate the sequence of the process described.

3. To be effective, graphics must be well integrated with the text, as in this example. After the reader is referred to a flow chart (Figure 1), each sentence of the narrative that follows corresponds to a specific site on that figure. Connect each sentence with its appropriate site on Figure 1. Compare the effectiveness of this flow chart with that in Figure 2. How are these two figures related to the subject of the description? Do they require different degrees of explanation to be understood by the reader?

Application

Design a flow chart for a process with which you are familiar and which lends itself to graphic representation. Once you have finished your sketch, write a text to explain it to your reader. ■

Processing Information in a Data Base

The following description of data base comes from *IMS/VS Version 1 Application Programming: Designing and Coding.* Part of IBM's product support literature, this manual is written to help programmers at installations operating under IMS/VS (an acronym for "Information Management System/Virtual Storage") to develop programs to manage large quantities of information. Central to problems of information management is the concept of data base described here. To understand the manual, the programmer must know one computer language, COBOL, but needs no experience with information management systems. Even though this description is aimed at a moderately specialized audience familiar with computer programming, it uses a number of techniques that make the description easy to understand. Most obvious is the extended example of a medical clinic's record management and design.

1   This section describes what makes storing data in a data base different from other ways of storing data.

Comparing Ways to Store Data

2 The advantage of storing and processing data in a data base is that all of the data appears only once, and that each program has to process only the data that it needs. One way to understand this is to compare three ways of storing data: in separate files, in a combined file, and in a data base.

3 *Storing Data in Separate Files*   If you keep separate files of data for each part of your organization, you can make sure that each program uses only the data it needs, but you have to store a lot of the data in several places at once. The problem with this is that redundant data takes up space that could be used for something else.

4   For example, suppose that a medical clinic keeps separate files for each of its departments, such as the clinic department, the accounting department, and the ophthalmology department.

- The clinic department keeps data about each patient that visits the clinic. For each patient, the clinic department needs to keep this information:

  —*The patient's identification number*
  —*The patient's name*
  —*The patient's address*
  —*The patient's illnesses*
  —*The date of each illness*
  —*The date that the patient came to the clinic for treatment*
  —*The treatment that was given for each illness*
  —*The doctor that prescribed the treatment*
  —*The charge for the treatment*

- The accounting department also keeps information about each patient. The information that the accounting department might keep for each patient is:

  —*The patient's identification number*
  —*The patient's name*
  —*The patient's address*
  —*The charge for the treatment*
  —*The amount of the patient's payments*

- The information that the ophthalmology department might keep for each of its patients is:

  —*The patient's identification number*
  —*The patient's name*
  —*The patient's address*
  —*The patient's illnesses that relate to ophthalmology*
  —*The date of each illness*
  —*The names of the members in the patient's household*
  —*The relationship between the patient and each household member*

5   If each of these departments keeps separate files, each department uses only the data that it needs, but a lot of data is redundant. For example, every department in the clinic uses at least the patient's number, name, and address. Updating the data is also a problem because if several departments change the same piece of data, you have to update the data in several places. Because of this, it's difficult to keep the data in each department's files current. There's a danger of having current data in one department and "old" data in another.

6   ***Storing Data in a Combined File***   Another way to store data is to combine all of the files into one file for all of the departments at the clinic to use. In the medical example, the patient record that would be used by each department would contain these fields:

- The patient's identification number
- The patient's name
- The patient's address
- The patient's illnesses
- The date of each illness
- The date that the patient came to the clinic for treatment
- The treatment that was given for each illness
- The doctor that prescribed the treatment
- The charge for the treatment
- The amount of the patient's payments
- The names of the members in the patient's household
- The relationship between the patient and each household member

7   Using a combined file solves the updating problem because all of the data is in one place, but it creates a new problem: the programs that process this data have to access the entire data base record to get to the part that they need. For example, to process only the patient's number, charges, and payments, an accounting program has to access all of the other fields as well. In addition, changing the format of any of the fields within the patient's record affects all of the application programs, not just the programs that use that field. Using combined files can also involve security risks, since all of the programs have access to all of the fields in a record.   ■

8   *Storing Data in a Data Base*   Storing data in a data base gives you the advantages of separate files and combined files: all of the data appears only once, and each program accesses only the data that it needs. This means that:

- When you update a field, you only have to update it in one place.
- Since you store each piece of information only in one place, you can't have an updated version of the information in one place and an out-of-date version of the information in another place.
- Each program accesses only the data it needs.
- You can keep programs from accessing private information.

9   In addition, storing data in a data base has two advantages that neither of the other ways has:

- If you change the format of part of a data base record, the change doesn't affect the programs that don't use the changed information.
- Programs aren't affected by how the data is stored.

10   Because the program is independent of the physical data, a data base can store all of the data only once and yet make it possible for each program to use only the data that it needs. In a data base, what the data looks like when it's stored, and what it looks like to an application program are two different things.

Figure 1

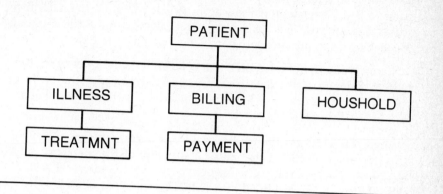

What the Data Looks Like When It's Stored

11 In IMS/VS, a record is stored and accessed in a hierarchy. A hierarchy shows how each piece of data in a record relates to other pieces of data in the record. Figure 1 shows the hierarchy you could use to store the patient information described earlier in this chapter.

12   IMS/VS connects the pieces of information in a data base record by defining the relationships between the pieces of information that relate to the same subject. The result of this is a data base hierarchy. The hierarchy shows how each piece of information is related to other pieces of information in the record. The relationship between two pieces of information in the hierarchy means that one piece of information is either dependent on or equal to another piece of information.

13   In the medical data base, the data that you're keeping is information about a particular patient. Information that is not associated with a particular patient is meaningless. For example, keeping information about a treatment given for a particular illness is meaningless if the illness isn't associated with a patient. ILLNESS, TREATMNT, BILLING, PAYMENT, and HOUSHOLD must always be associated with one of the clinic's patients to be meaningful information.

14   There are five kinds of information you're keeping about each patient. The information about the patient's illnesses, billings, and household depends directly on the patient. The information about the patient's treatments and the patient's payments depends respectively on the patient's illnesses and the patient's payments as well.

15   Each of the pieces of data represented in Figure 1 is called a segment in the hierarchy. A segment is the smallest unit of data that an application program can retrieve from the data base. Each segment contains one or more fields of information. The PATIENT segment, for example, contains all of the information that relates strictly to the patient: the patient's identification number, the patient's name, and the patient's address.

What the Data Looks Like to Your Program

16 IMS/VS uses two kinds of control blocks to make it possible for application programs to be independent of the way in which you store the data in the data base. One control block defines the physical structure of the data base; another defines an application program's view of the data base:

- A data base description, or DBD, is a control block that describes the physical structure of the data base. The DBD also defines the appearance and contents, or fields, that make up each of the segment types in the data base.

  For example, the DBD for the medical data base hierarchy shown in Figure 1 would describe to IMS/VS the physical structure of the hierarchy, and it would describe each of the six segment types in the hierarchy: PATIENT, ILLNESS, TREATMNT, BILLING, PAYMENT, and HOUSHOLD.

- A data base program communication block, or DB PCB, in turn, defines an application program's view of the data base. An application program often needs to process only some of the segments in a data base. A PCB defines which of the segments in the data base the program is allowed to access. The program is "sensitive" to the segments that it's allowed to access. The data structures that are available to the program contain only segments that the program is sensitive to.

For example, an accounting program that calculates and prints bills for the clinic's patients would need only the PATIENT, BILLING, and PAYMENT segments. You could define the data structure shown in Figure 2 in a DB PCB for this program.

Figure 2

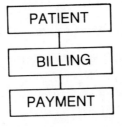

17   A program that updates the data base with information on patients' illnesses and treatments, on the other hand, would need to process the PATIENT, ILLNESS, and TREATMNT segments. You could define the data structure shown in Figure 3 for this program.

Figure 3

18   Sometimes a program needs to process all of the segments in the data base. When this is true, the program's view of the data base as defined in the DB PCB is the same as the DL/I hierarchy that's defined in the DBD.

19   Each DB PCB defines a way in which the application program views and processes the data base. The DB PCB also tells IMS/VS how the program is allowed to process the segments in the data structure—whether the program can only read them, or whether it can update segments as well.

20   A program specification block, or PSB, contains the DB PCBs for a particular application program. A program may use only one DB PCB—which means it processes only one data structure—or it may use several DB PCBs, one for each data structure. There is one PSB for each application program.

21   Since an application program processes only the segments in a data base that it requires, if you change the format of a segment that a program doesn't process, you don't have to change the program. A program is affected only by the segments that it accesses. In addition to being sensitive to only certain segments in a data base, a program can also be sensitive to only certain fields within a segment. This is called field level sensitivity. If you change a segment that the program isn't sensitive to, it doesn't affect the program. In the same way, if you change a field that the program isn't sensitive to, it doesn't affect the program.

How You Process a Data Base Record

22   A data base record is a root segment occurrence and all of its dependents. In the medical example, a data base record is all of the information about one patient. The PATIENT segment in the medical data base is called the root segment. The segments below the root segment are called dependents, or children, of the root. For example, ILLNESS, BILLING, and HOUSHOLD are all children of PATIENT. ILLNESS, BILLING, and HOUSHOLD are called direct dependents of PATIENT; TREATMNT and PAYMENT are also dependents of PATIENT, but they are not direct because they are at a lower level in the hierarchy.

23   Each data base record has only one root segment occurrence, but it may have several occurrences at lower levels. For example, the data base record

for a patient contains only one occurrence of the PATIENT segment type, but it may contain several ILLNESS and TREATMNT segment occurrences for that patient.

24   To process the information in the data base, your application program communicates with IMS/VS in three ways:

- Passing control: IMS/VS passes control to your application program through an entry statement in your program. Your program returns control to IMS/VS when it has finished its processing.
- Communicating processing requests: Your program communicates processing requests to IMS/VS by issuing calls to Data Language I, or DL/I. DL/I is an access method that handles the data in the data base.
- Exchanging information with DL/I: Your program exchanges information with DL/I through two areas in your program. First, DL/I reports the results of your calls in the DB PCB. Your program builds a mask of the DB PCB and uses this mask to check the results of the calls. Second, when you request a segment from the data base, DL/I returns the segment to your I/O area. When you want to update a segment in the data base, you place the new value of the segment in the I/O area.

25   An application program can read and update a data base. When you update a data base, you can replace segments, delete segments, or add segments. You indicate to DL/I the segment you want to process, and whether you want to read or update it, in a DL/I call. ∎

Questions for Analysis and Discussion

1. How effective is the example of the medical clinic's records and other elements of the description in making concrete the abstract concept of data base? Can you think of other examples that would also be effective?

2. This manual is designed for someone new to information management systems. How does the format—headings, spacing, and other graphic elements—organize the material to facilitate the reader's progression through the text?

3. Compare the rather informal tone of this selection with the more formal tone of the description of oil shale processing (pp. 66–68) or of the descriptions of Celanese wastewater treatment (pp. 60–61 and 62–65). Is the informal tone of this description of data base appropriate for the manual's audience? Why or why not?

4. Both this description of data base and the selection "What Is a Computer?" (pp. 24–25) concern computers. Both use a conversational tone, examples, and other techniques to help the reader understand the subject. However, the audience for "Data Base" is quite different from that of "What Is a Computer?" as are their purposes. What are these differences in audience and purpose, and how are they manifested in each selection?

Application

Select an abstract concept or process with which you are familiar and write a description that centers around an extended example. Use the example to make the abstraction concrete and familiar.

3

INSTRUCTIONS

Explaining or understanding how to do something is part of our common experience. Every day we give or follow instructions—while preparing a recipe, repairing the garbage disposal, finding an address. Technical instructions, procedural manuals, and users' guides are simply the formal equivalent of this day-to-day activity. And the same basic requirements of clarity, precision, and accuracy apply for all instructions, whether we are telling children how to brush their teeth or writing a program to send a manned spacecraft safely into orbit. In each case, a flaw in the instructions can have unexpected or undesired results in the performance of the task.

The success of a set of instructions depends on the performer's ability to follow the instructions easily and without confusion. In preparing instructions, procedural manuals, or guidelines, the writer should subordinate everything else to this purpose. Organization, word choice, layout, graphics, even the color and size of print are chosen to make the performance of the task as clear and simple as possible.

A set of instructions can be organized in two ways: chronologically and topically. We're familiar with many sets of instructions organized *chronologically,* that is, as a series of steps arranged in the order of their performance. However, many other instructions do not lend themselves to simple chronological ordering because the tasks involved cannot easily be divided into a sequential series of steps. And so these are organized according to topic, as is often the case, for example, with general guidelines.

Some sets of instructions or procedural manuals use a combination of chronological and topical organization. For instance, the instructions for changing the oil in an engine (pp. 83–88) are organized in a step-by-step sequence. However, the manual from which these instructions were taken uses an overall topical pattern of organization, based on various aspects of machinery maintenance.

Because the purpose of chronological instructions is so specific, limited, and easy to define, such instructions are probably the easiest form of technical writing to produce—if you are attentive to detail. The organization of time-ordered instructions is imposed by the order in which each step must be performed. The main consideration is to provide the performer with everything—not more, not less—required to complete the task successfully.

Chronological instructions usually begin with introductory information that prepares the reader to perform the procedure. This information may include:

- A detailed description of the task to be performed and fundamental principles of operation—that is, what is to be done and why it is to be done in this manner. An overview of the entire procedure and its major steps may also be given. If the reader can understand the logic of the procedure and see it as a whole, he or she is likely to find it easier to follow subsequent details.
- A complete description of all tools, equipment, ingredients, or special conditions needed to perform the task. This information allows the gathering of all necessary materials at the start and prevents the reader from realizing, halfway through the task, that it can't be completed without a particular tool or ingredient. All cooks know the dismay of realizing that "This recipe calls for two cups of milk, and there's not a drop left in the house!"
- Any necessary cautions about possible hazards or dangers to the reader. These warnings will often be highlighted by color, size of print, or immediately recognizable symbols—the skull and crossbones, a stop sign, the radioactive warning logo—to ensure that the reader heeds the cautionary note.

The body of chronological instructions consists of a step-by-step description of all tasks to be done in the sequence in which they must be performed. Here, instructions are organized in the same manner as process descriptions (see pp. 41–42 of Chapter 2): An entire procedure is divided into a series of sequential actions. But while process descriptions commonly use narrative to tell what is happening, chronological instructions usually use the imperative to tell how to do it. Instructions address the reader directly by using the active voice and imperative mood (for example, "Catch all the oil in a bucket," rather than "All the oil is then caught in a bucket").

Good instructions are usually written so that the reader can perform the process while reading the manual, pamphlet, or reference guide. For this reason, instructions should look as uncluttered and organized as possible. Major steps may be subdivided into short tasks, each of which can be done within a brief span of time. Sometimes these are presented in tabular form, as in the recipe for beef stew (pp. 91–92). Numerals can also help the performer keep track of progress. Other layout features—such as white space, color choice, and typeface—can make the instructions more readable. The McLane Edger and Trimmer Owner's Manual, pp. 98–108, is an example of a well laid-out set of instructions.

Many instructional documents, such as guidelines or procedural manuals, are more easily organized topically than chronologically. In certain circumstances, the topical breakdown of these instructions may be mandated, but it is usually the writer's responsibility to decide what topics need to be discussed. Sometimes the writer will employ an arbitrary pattern of organization, using letters, numerals, or outline format to make topic division clear. In other cases, the nature of the information to be transmitted will provide the writer with a logical arrangement of topics. For instance, the writer of the OSHA pamphlet on carcinogens in the workplace (pp. 115–18) was guided by federal regulations concerning this problem. And these regulations themselves are organized by topic, according to employers' concerns about hazardous material.

The choice between a chronological or a topical pattern of organization is often determined by the level of the material. The more general the material, the more likely the writer will use a topical organization. Conversely, if the material involves specific, concrete applications, a chronological organization is the probable choice. Let's look at an example. "Self-help" books range from *How to Lose X Pounds in Y Days* to *How to Live a Fulfilled Life.* To be a successful book, *How to Lose* must describe very specific actions and provide daily diets and guidance—and so it likely will have a chronological organization. On the other hand, the subject of *How to Live* is so abstract—and the definition of "fulfilled life" is itself so problematic—that a chronological plan of progress would be difficult to develop. In that case, a topical organization would be the best.

The clear purpose and intended audience of instructions determine the language used and amount of detail included in them. For example, the instructions on measuring thresholds of taste (pp. 110–13) and those for preparing beef stew (pp. 91–92) are obviously written for readers with very different degrees of technical knowledge who need to perform very different kinds of tasks. When writing for the inexperienced reader, the writer defines key terms and avoids overly theoretical material. The technically experienced reader, however, often wants the theoretical background because the instructions may need to be adapted to a range of circumstances. A surgeon who is studying a new surgical procedure, for instance, wants to know *how* to do the procedure, and is equally concerned with *when* and *why* the procedure can and should be done.

Instructions should also be as complete as necessary. Failing to assess correctly how much the reader knows about the task, the inexperienced writer often gives too much or too little information. The writer may assume that the reader knows which way to turn the screw to get it off, or that the chemicals used in the task can be toxic if the workplace is not properly ventilated.

But if the reader doesn't know this information, the instructions are confusing—or even dangerous. On the other hand, telling the reader too much can be as confusing as saying too little. In some cases, an abundance of detail may be necessary to ensure against every eventuality—especially when legal consequences are present, as in contracts, insurance policies, or wills. But in most instances, too much detail interferes with effective communication. A rule to follow: Grant the reader a certain amount of common (and sometimes even uncommon) sense.

How to Change Oil

These instructions for changing oil and oil filters come from *Machinery Maintenance,* a manual produced by the John Deere Service Training Department of Deere & Company. The manual is designed to introduce the fundamentals of machinery maintenance to an international audience. Because the instructions are written for readers with minimal mechanical experience, they are thorough and take little knowledge for granted. Also, because the manual will be translated into various languages, sentence structure and word choice are basic and the use of local idiom is avoided. Deere & Company have also found that illustrations greatly simplify the task of translating technical literature.

HOW TO CHANGE OIL

When oil begins to wear out, it can't protect your engine.

The oil and oil filter must be changed when recommended in your operator's manual. Waiting longer causes wear and may even ruin an engine.

DRAIN THE OLD OIL

Be sure you have:

- *Enough of the recommended oil*
- *The recommended oil filter*
- *Drain plug wrench*
- *Drain pan*
- *A clean cloth*

Before changing oil, run the engine until it reaches normal operating temperature. Running the engine mixes the dirt and sludge in the crankcase with the oil so it can all run out together.

1. Stop the engine and remove the crankcase drain plug or plugs.

2. Catch all the oil in a bucket. Dispose of the old oil so it doesn't pollute. Some refining companies will buy the used oil.

CAUTION: Never operate the engine without oil in the crank-case.

1. Let all the oil drain from the crankcase.

2. If the drain plug is magnetic, clean off pieces of metal carefully. Do not damage the threads.

3. Wipe the drain plugs clean and put them back in the crankcase.

4. Wipe the oil and grime off the crankcase so it can radiate heat.

REMOVE THE OLD OIL FILTER

There are three main kinds of oil filters:

- *Through bolt*
- *Internal*
- *Spin on*

Before removing the filter:

1. **Wipe dirt and grease from the filter and filter mounting.**

2. **Remove the drain plug and catch the oil in a container.**

3. **Remove the filter.**

4. **Always wipe off the filter mount with a clean cloth. Careful, don't leave lint on the mount.**

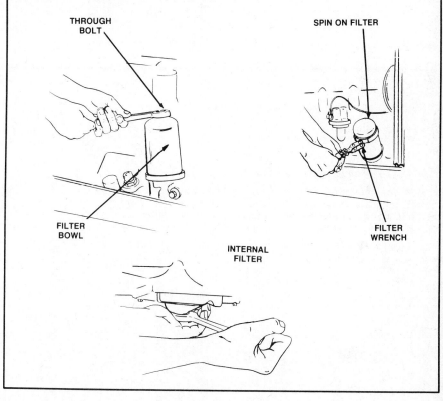

THE FILTER

Most engines have spin on oil filters you can discard. Some of these filters have a relief valve built into the filter itself. Make sure you use the right filter.

Some have filter elements that are disposable, but the housings remain with the machine. Wipe these housings clean with a cloth before inserting a new element. Internal filters have a removable cap you must clean and replace.

MOUNT THE NEW FILTER

After the oil is drained, the drain plugs replaced, and you have done your cleanup, mount a new oil filter on the engine.

1. **Wipe off filter base. Leave no lint.**

2. **Use a new filter gasket.**

3. **Spread a thin coat of new oil on the filter gasket.**

4. **Mount the new filter. If it is a spin on filter, turn it on by hand until the gasket touches the mounting face. Then turn it another half turn until it is snug. Do not overtighten it or you may crush the gasket and cause a leak.**

5. **If you have a through bolt or internal oil filter, insert the correct new filter using a new, lightly oiled gasket. Snug it up, but do not overtighten or the gasket will be crushed and leak.**

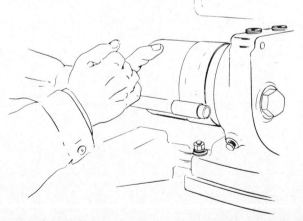

CLEAN OIL CANS AND SPOUTS

Before refilling the crankcase, wipe the dirt off spouts and cans.
Dirt can be carried into the engine with the oil.

REFILL THE CRANKCASE WITH OIL

After the new oil filter is on and the drain plugs are back in place:

1. Check your operator's manual to be sure you have the correct amount and kind of oil.

2. Open the crankcase filler and pour in the correct amount of oil. Your operator's manual will tell you where to find the filler.

3. Do not overfill the crankcase. Overfilling causes oil to foam and leak.

4. Let the oil drain down for a few minutes.

5. Start the engine and let it run slowly until oil pressure reaches normal.

6. Do not run the engine fast until the oil pressure is normal.

7. Stop the engine and check around the oil filter and drain plugs for leaks. Tighten the filter and drain plugs, if they leak.

8. Check the oil level again. Add oil if the level is too low.

9. Wipe spilled oil off the engine.

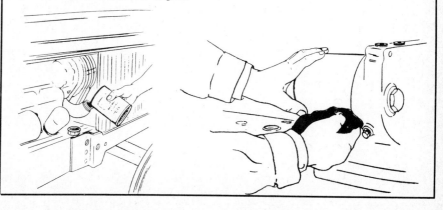

RECORD ENGINE HOURMETER READING

Always write down the date and engine hourmeter reading when you change the oil and oil filter. Then you can figure out when the next service is required. Servicing the engine on time and keeping good records shows good judgment. It saves time and money and reduces the chance of machine failure.

Questions for Analysis and Discussion

1. The audience and purpose of this manual require that the content and writing style be as simple as possible. Point to specific examples that show how this simplicity has been achieved.

2. How are these instructions organized? You might want to construct a brief outline of the reading to help determine its pattern of arrangement. How logical do you find the organization? How helpful are the headings and subheadings in revealing the organization to the reader?

3. These instructions, like many others, use illustrations to help the performer successfully complete each task. How helpful are the illustrations in clarifying the process? Why do you think human figures were included in the drawings? What characteristics do the people portrayed possess? Are these characteristics determined by the intended international audience?

4. Many of you change the oil in your car or are at least familiar with the process. Evaluate the usefulness of these instructions based on your experience or knowledge. In adapting these instructions to your car, what details would you add? Do these instructions leave any questions unanswered?

Application

Write a thorough set of instructions for a fairly simple process so that it can be understood by a reader with little specialized knowledge. Consider carefully the reader's needs as you choose the language, format, and amount of detail and explanation your instructions will contain.

Beef Stew

A recipe is one of the most basic and familiar kinds of instructions. Everyone who has encountered vague or confusing directions knows how frustrating that can be. But what makes a recipe complete depends on the individual using it. Some readers can apply such directions as "Season to taste"; others need exact measurements. Preparation of food for large groups, such as the military, makes the need for standardization greater, as this recipe from the Armed Forces' *Recipe Service* shows. The problems of military cooking are compounded by the fact that some kitchen personnel may be inexperienced in preparing food. Thus these instructions leave no detail unspecified, and list precisely and completely all ingredients and methods necessary to prepare the Army's beef stew.

YIELD: 100 Portions (2 Pans)   EACH PORTION: 1 1/4 Cups

PAN SIZE: 18 by 24-inch Roasting Pan

INGREDIENTS	WEIGHTS	MEASURES		METHOD
Beef, diced, thawed	30 lb	....	....	1. Dredge beef in mixture of flour, salt, pepper, and garlic; shake off excess.
Flour, wheat, general purpose, sifted	8 oz	2 cups	....	
Salt	5 oz	1/2 cup	....	
Pepper black...............	....	2 tbsp	....	
Garlic	....	1 2/3 tbsp	....	
Shortening, melted or salad oil	1 lb	2 cups	....	2. Brown beef in hot shortening or salad oil.

(OVER)

INGREDIENTS	WEIGHTS	MEASURES		METHOD
Water, hot	....	2 1/2 gal	....	3. Add water, tomatoes, thyme, and bay leaves to meat. Cover; simmer 2 hours.
Tomatoes canned, crushed	6 lb 6 oz	3 qt (1-No. 10 cn)	....	
Thyme, ground	....	1 tbsp	....	
Bay leaves, whole	....	4 leaves	....	
Carrots, fresh, 1/2-inch rings	8 lb	3/4 gal ..	....	4. Add carrots and rutabagas to meat mixture. Cover; simmer 15 minutes.
Rutabagas, fresh, diced	2 lb	2 qt	....	
Celery, fresh, cut in 1-inch pieces	4 lb	1 gal	....	5. Add celery and onions, Simmer 10 minutes.
Onions, dry, cut in quarters	3 lb	2 1/2 qt	....	

(CONTINUED)

INGREDIENTS	WEIGHTS	MEASURES		METHOD
Potatoes, white fresh, peeled cut in 1 to 1 1/2- inch pieces	8 lb	1 1/2 gal	....	6. Add potatoes and salt. Stir to mix. Cover; simmer 20 minutes or until vegetables are tender.
Salt	2 oz	3 tbsp	....	
Flour, wheat general purpose, shifted	1 lb 2 oz	4 1/2 cups	....	7. Thicken gravy, if necessary. Combine flour and water. Add to stew while stirring. Cook 5 minutes or until thickened.
Water, cold	....	1 1/2 qt.	....	

(OVER)

NOTE 1. In Step 1, 30 lb beef, pot roast, diced in 1 to 1 1/2-inch pieces may be used. Trim beef to remove excess fat and gristle.

2. In Step 4, 9 lb 12 oz fresh carrots A.P. will yield 8 lb carrot rings and 2 lb 6 oz fresh rutabagas will yield 2 lb rutabagas.

3. In Step 4, if rutabagas are not available, increase potatoes, fresh, white in Step 6 to 10 lb (12 lb 5 oz A.P.).

4. In Step 5, 5 lb 8 oz fresh celery A.P. will yield 4 lb celery cut into 1-inch pieces and 3 lb 5 oz dry onions A.P. will yield 3 lb onions cut in quarters.

5. In Step 5, 6 oz (2 cups) dehydrated onions may be used. See Recipe Card A-11.

6. In Step 6, 9 lb 12 oz fresh white potatoes A.P. will yield 8 lb potatoes cut into 1 to 1 1/2-inch pieces.

7. In Steps 2 through 7, a steam-jacketed kettle or 2 roasting pans (18 by 24 inches) on top of range, or 350°F. oven may be used to brown and cook meat and vegetable mixture.

8. In Steps 4 through 6, vegetables may be cooked separately and then added to browned meat.

9. Other sizes and types of pans may be used. See Recipe Card A-25.

Questions for Analysis and Discussion

1. This recipe is designed, in part, to be used as a training aid in military food-service schools. What elements of the recipe suggest that it has been written for a relatively inexperienced cook?

2. Why are measurements for some ingredients given in both weight and volume?

3. Compare this recipe with the instructions for changing oil (pp. 83–88). What features do these two sets of instructions have in common?

Application

Discuss the advantages of the gridiron format of this recipe. Select a recipe or a similar simple process with which you are familiar and rewrite it to conform to this pattern.

Mavis R. Cranston
Datarite Corporation

Joining a Credit Union

The following brief instructions for joining a credit union are presented as a memorandum. Succinct and well-organized, the memo focuses on the information new employees may need. It also anticipates possible questions the reader might raise by supplying that information in a very specific and detailed manner.

To: All New Employees
From: Mavis R. Cranston
Re: Colorado Central Credit Union

1 The Colorado State Bank Commissioner has approved Datarite Data Systems as an employee group, permitting all employees to become members of the Colorado Central Credit Union.

2 Data Systems payroll has made arrangements to allow payroll deductions for the Colorado Central Credit Union. If you wish to join the credit union:

1. Get an application from Personnel.

2. Complete both sides of the application. Whenever a form requires the name of the credit union of which you are a member, insert "Datarite Corporation."

3. Return the completed application to the credit union in the prepaid postage envelope with your check or money order in the amount of $6.00. One dollar is for membership dues; the remaining $5.00 opens your account.

4. You will receive a letter of acknowledgement listing your account number and personal representative from Colorado Central Credit Union.

5. Regular payroll deductions are then remitted to Central in your account.

6. You will receive monthly statements from the credit union, along with information about their services.

Mavis R. Cranston
Manager, Personnel

Questions for Analysis
and Discussion

1. How does the first paragraph briefly establish the memo's context and purpose?

2. Why is a numbered list of sentences an effective way to convey the information these instructions detail?

3. What is the difference between the first three items in the list of instructions and the last three? How is this difference reflected in sentence construction?

Application

Using memo format, write a brief set of instructions explaining a common activity at your school, such as obtaining athletic tickets, getting a parking permit, or logging on to the computer.

Owner's Handbook
Lawn Edger and Trimmer

The McLane lawn edger and trimmer is designed for both grounds maintenance professionals and for homeowners. The following handbook instructs
the owner of the edger how to assemble, operate, and care for the equipment.
Outstanding features of these instructions include the emphasis on user safety, the use of graphics, and the comprehensiveness with which this fairly simple machine is treated. The manual also contains a troubleshooting guide, a
special section found in many sets of instructions which is organized to allow
quick diagnosis of problems and easy access to possible solutions. This section is usually designed as a list or table; the reader who needs additional
information is referred to an appropriate part of the manual.

McLANE

Owner's Handbook
Lawn Edger and Trimmer

Models   100-2R6   800-3RB
             100-2R7   100-3FR
             800-3RP   4-K
             100-3R   4-FK

Safety Rules

⚠ **Improper use of the edger can result in injury. To reduce this possibility, give complete and undivided attention to the job at hand, and follow these safety precautions.**

Training
1. Read the Operating and Service Instruction Manual carefully. Be thoroughly familiar with the controls and proper use of the equipment.
2. Never allow children to operate edger-trimmer.
3. Keep area of operation clear of all persons, particularly small children and pets.

Preparation
1. Thoroughly inspect the area where edger-trimmer is to be used, and remove all stones, sticks, wire, bones and other foreign objects which might be picked up and thrown.
2. Do not operate edger-trimmer when barefoot or wearing open sandals. Always wear substantial foot wear, and pants or slacks that cover your legs when operating edger.
3. Check fuel before starting engine. Do not fill gasoline tank indoors, when engine is running, or while engine is still hot. Wipe off any spilled gasoline before starting engine.

4. Disengage cutterhead drive before starting engine.
5. Do not use edger-trimmer unless blade guard, belt guards and debris guard are properly in place.
6. Edge or trim only in daylight or in good artificial light.
7. Never operate equipment in wet grass. Always be sure of your footing: keep a firm hold on the handle and walk, never run.

Operation
1. Start the engine carefully. Keep hands and feet well away from blade.
2. Do not change engine governor setting or overspeed engine.
3. Wear safety glasses when operating edger-trimmer.
4. Keep clear of discharge opening at all times. Never direct discharge of any material toward bystanders nor allow anyone near machine while it is in operation.
5. Stop blade when crossing gravel drive, walks or roads.

---

6. After striking a foreign object, stop the engine and inspect edger-trimmer for damage: repair damage before starting engine.
7. Avoid loss of control or tipping on slopes and sharp corners by reducing speed.
8. If edger-trimmer should start to vibrate abnormally, stop the engine and check for the cause. Vibration is generally a warning of trouble.
9. Stop the engine whenever you leave the edger-trimmer, and when making repairs or inspections.
10. When repairing or inspecting, make certain blade and all moving parts have stopped. Disconnect spark plug wire and keep wire away from plug to prevent accidental starting.
11. Do not run engine indoors.
12. Shut engine off and wait until blade comes to a complete stop before removing grass that may clog blade guard.
13. Watch out for traffic when working near roadways.
14. Stay alert for uneven sidewalks, holes in terrain or other hidden hazards when using edger-trimmer. Always push slowly over rough ground.
15. Remember, never operate edger-trimmer unless blade guard, belt guards and debris guard are properly in place.
16. Disengage power to cutterhead and stop engine before leaving edger-trimmer, before making repairs, inspections or adjustments, and when transporting or not in use.

Maintenance and Storage

1. Check blade mounting nut frequently for proper tightness.
2. Keep all nuts, bolts and screws tight to be sure equipment is in safe working condition.
3. Never store edger-trimmer with gasoline in the tank in a building where fumes may reach an open flame or spark. Be sure engine has cooled before storing in any enclosure.
4. To reduce fire hazard, keep engine free of grass, leaves or excessive grease.
5. After operating engine, never touch exhaust muffler until it has cooled completely.
6. Keep edger-trimmer in good operating condition and keep safety devices in place.

WARNING TO PURCHASERS OF INTERNAL COMBUSTION ENGINE EQUIPPED MACHINERY OR DEVICES

This equipment does not include a spark arrestor muffler.
If this equipment is used on any forest-covered land, brush-covered land or grass-covered land in the State of California, the law requires that a spark arrestor be installed and in effective working order before using this equipment on such land. The spark arrestor must be attached to the exhaust system and comply with Section 4442 of the California Public Resources Code.
If this equipment is used in a state or national park or campground, a U.S.D.A. Forestry Department approved spark arrestor muffler must be utilized.

Lawn Edger and Trimmer Features

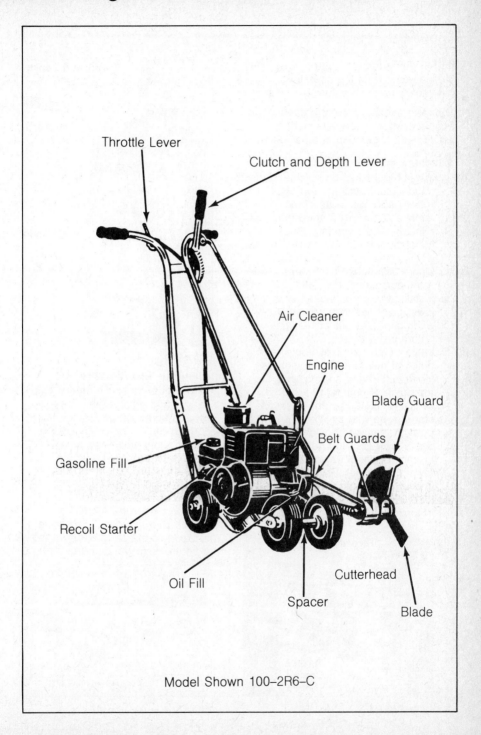

Throttle Lever

Clutch and Depth Lever

Air Cleaner

Engine

Blade Guard

Belt Guards

Gasoline Fill

Recoil Starter

Oil Fill

Spacer

Cutterhead

Blade

Model Shown 100–2R6–C

Assembly

NOTE: Reference to Left or Right side of machine is from the operator's position at the handle, facing forward.

The lawn edger, except the handle and throttle control, is fully assembled, packed and shipped in one container.

Attaching the Handle

1. Install handlebars on outside frame of uprights. Install the top bolts first with nuts to the inside. Then remove center bolts holding spacer panel between uprights. Pivot handlebars to upright positions and install center and bottom bolts. Leave bolts loose until Steps 2 and 3 are completed. (Fig. 3)
2. Install control rod into upper and lower pivot bearings and secure with cotter pins. Oil pivots lightly before installing rod.
3. Move clutch-depth lever into rear notch of quadrant on handlebars. Move cutterhead angling lever to top notch of quadrant.
4. Check for clutch by pulling recoil starter rope with throttle in *off* position, if blade should turn when depth lever is in top notch then loosen handle bolts and pull towards operator. Snug bolts and check for clutch again. Make sure blade does not turn when lever is in top notch.

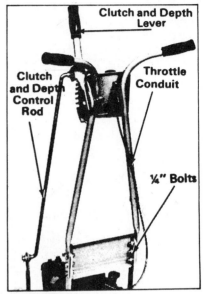

Figure 3

Installing the Throttle Wire

1. Remove thumb screw on top of air cleaner and remove air cleaner from carburetor (See Fig. 1).
2. Route throttle wire conduit below spacer panel between frame uprights before installing handlebars. Install throttle wire in tophole (from inner side, toward engine) of vertical part of control lever on carburetor (Fig. 1).

NOTE: Governor spring must be hooked into second hole from top of control lever on carburetor (Fig. 2)

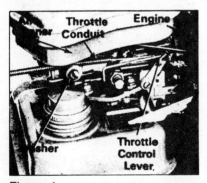

Figure 1

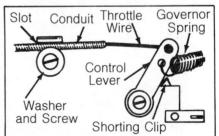

Figure 2

Hints for Best Performance

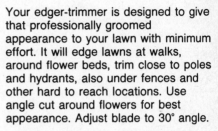

3. Loosen screw and install conduit in slot behind washer. Leave screw loose until handlebars are installed.
4. After handlebars are installed, adjust throttle control by moving throttle lever on the control panel fully to the rear "OFF" position.
5. Slide conduit and wire forward in slot behind washer until control lever strikes shorting clip. Tighten screw firmly to hold conduit.
6. Replace air cleaner.

Your edger-trimmer is designed to give that professionally groomed appearance to your lawn with minimum effort. It will edge lawns at walks, around flower beds, trim close to poles and hydrants, also under fences and other hard to reach locations. Use angle cut around flowers for best appearance. Adjust blade to 30° angle.

For best results adjust the cutting depth of the blade to within ½" to 1" deep, with a new blade usually in the third notch of the adjustment bracket at handle.

Too deep a cut will put unnecessary stress on engine and accelerate blade wear. Cutting too shallow allows rapid regrowth of roots.

Make sure to re-adjust cutting depth as blade begins to wear at tips.

Avoid edging or trimming in wet grass as it may clog up in the blade guard and cause splattering. The general appearance will be much better when grass is dry.

Always push edger at moderate walking speed—at this speed your edger-trimmer is designed to cut more than 100 ft. per minute (For idle adjustment see engine maintenance section). Four wheel design of edger permits easy and stable edging riding securely on sidewalk or cement surfaces.

Around specially shaped flower beds or curved sections follow these simple steps. It will be easier to cut around corners when front wheels are placed together at right side of edger. Before starting engine, remove wheel spacer clip on front axle, slide wheels to side away from blade, replace spacer.

After starting engine, set blade depth and by slightly lifting rear wheels, guide edger around corners and odd angles.

Operation

NOTE: Because the rotating blade can throw dirt or debris, wear safety glasses when operating the edger.

Adjusting Cutterhead Angle Before starting engine adjust cutterhead angle. The cutterhead angling lever, located on the left hand side of the frame (See Fig. 7), enables the operator to position the blade horizontally for trimming (Fig. 7) or choose several vertical positions for edging.

WARNING: Never adjust cutterhead angle when engine is running.

Clutch and Depth Lever Settings
Clutch and Depth lever (Fig. 4) is used to engage and disengage the drive belt which operates the blade. It is also used to regulate the depth of edging and height of trimming.

Throttle Lever
1. The throttle lever is located at the right of the control panel and is used to regulate engine speed. (See Fig. 4)
2. Use FAST position when edging or trimming. Use SLOW position to idle the engine.

Starting the Edger
1. Be sure the clutch lever is in the rear DISENGAGED position before starting engine.
2. Move throttle lever to choke position. Pull rewind starter handle rapidly. Repeat if necessary. Move throttle up to running position.
3. Move Clutch and Depth lever forward to engage drive belt and blade. Notches in base of lever permit adjusting depth of cut for edging and height of trimming.
4. To stop edger, disengage Clutch & Depth lever and move throttle to stop position.

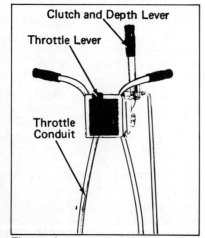

Figure 4

Maintenance

Wheel Positioning

1. The wheel adjusting spacer should be centered between the wheels when edging in the vertical position. Wheels must be in the dual position to the right when edging with cutterhead in an angle position or trimming (horizontal position). (See Fig. 5)

   WARNING: Do not change wheel positions when engine is running.

2. Remove the spacer from between the wheels by lifting on spacer with open side down. Slide the left front wheel on the axle toward the right wheel and once again install the spacer on the axle between the frame and wheel. (Fig. 5)

   CAUTION: Before starting the engine be sure the wheels and cutterhead are positioned so the blade will not strike the wheels.

Figure 5

Blade Change

1. Place a ½" wrench on the inside of cutterhead shaft and a ¾" wrench on the blade nut, turn to loosen nut. Remove outer washer and old blade.
2. Replace blade with only McLane blade (available in 3-pack). Place outer washer the same direction as removed and tighten nut as snug as possible.

Figure 6

Lubrication

Fill the crankcase with oil before starting engine. Check oil lever regularly. Oil all exposed bearings, linkage and front wheels.

When changing oil, tip to drain while engine is warm.

Clean and re-oil air cleaner and element, clean cooling system, clean and reset spark plug gap and remove carbon deposits at regular intervals.

Fill fuel tank with clean, fresh regular grade gasoline. Do not mix oil with gasoline.

Figure 7

Lubricate four times a year at regular intervals. (See Fig. 7)

Lawn Edger and Trimmer Parts Illustration

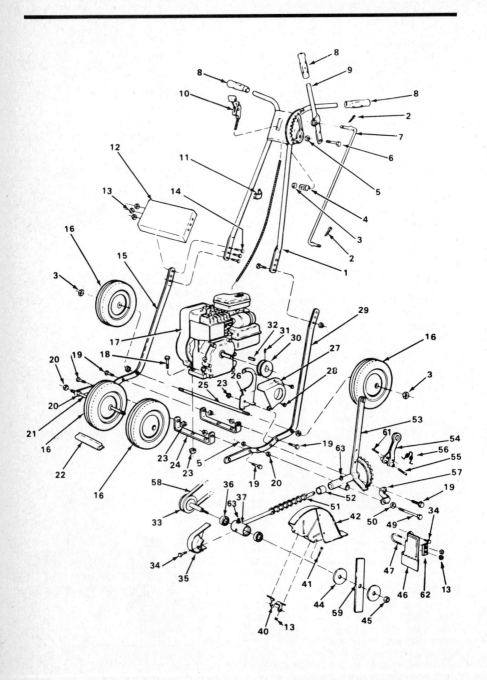

Lawn Edger and Trimmer Parts List

Order parts from McLane by specifying Model and Part Number

Ref. No.	Part No.	Description	Ref. No.	Part No.	Description
1	2001	Handle	37	2037	Mandrel, head
2	1011	*Pin, cotter, 1/8 × 1/2 (2 used)	40	2040	Clamp, blade guard
3	1008	*Nut Lock hex 7/16" 14 (4 used)	41	2041	*Bolt, 1/4-28 × 1-1/2 hex (2 used)
4	2004	Spring	42	2042	Guard, main
5	1007	Washer, plastic	44	2044	Washer, friction (2 used)
6	2006	Bolt, hex 7/16" 14 × 2"	45	2045	*Nut, 1/2-20
7	2007	Rod, Clutch and depth	46	2046	Deflector, dirt and debris
8	1009	Grip handle, used (3)	47	2047	U-Clamp
9	2009	Lever, clutch and depth	49	2049	Bolt, pivot hardened and ground
0	1013	Lever, remote throttle control	50	1120	*Flat Washer 7/16
1	2011	Clip, throttle cable	51	2051	Spring, head recoil
2	1014	Cross brace for handle	52	2052	Cup
3	1015	*Nut 1/4-28 hex (10 used)	53	2053	Body, cutterhead
4	1018	*Bolt 1/4-28 hex × 7/8 (6 used)	54	2054	Lever, angle shifting
5	2015	Frame, right hand	55	2055	Pin, 5/16 × 2
6	2016-6	Wheel, 6" (4 used)	56	2056	Spring
6	2016-7	Wheel, 7" (4 used)	57	2057	Brace, cutterhead
6	2016-8	Wheel, 8" (4 used)	58	2058	Drive Belt
7	2017	Engine	59	2059	Cutting Blade
8	1044	*Bolt, 5/16"-24 × 1-1/2" (4 used)	60	2060	Blade Wrench (accessory not illustrated)
9	1074	*Bolt, hex 5/16"-24 × 78" (5 used)	61	2061	1" Pin
0	2020	*Nut, hex 7/16-14 (3 used)	62	2062	U Clamp Spacer
1	2021-6, 7	Axle Front	63	2063	Grease Fitting
4	2021-8	Axle Front			
2	2022-6	Spacer Wheel			
2	2022-7	Spacer Wheel			
2	2022-8	Spacer Wheel			
3	1075	*Nut Hex 5/16-24 (9)			
2	2024-6, 7	Frame Crossmember (2 used)			
4	2024-8	Frame Crossmember (2 used)			
5	2025-6, 7	Axle, rear			
5	2025-8	Axle, rear			
6	2026	Bracket, engine pulley guard			
7	2027	Guard, engine pulley, back belt			
8	2028	*Screw, #10 × 1/2 sheet metal (2 used)			
9	2029-6, 7	Frame, left hand			
9	2029-8	Frame, left hand			
0	1063	Pulley, Engine			
1	1065	Set-screw 5/16"-24 × 5/16"			
2	1064	Key, 3/16 × 3/16 × 5/8			
3	2033	Pulley and shaft, cutterhead			
4	2034	*Screw, 1/4-20 × 1/2 hex			
5	2035	Guard, cutterhead front belt			
6	2036	Bearing, cutterhead (2)			

REPLACEMENT BEARING

Part No.	Bearing No.	No. Used
2009	1117	1 Lever, Clutch & Depth
2016-6	1117	8 Wheels
2016-7	1117	8 Wheels
2016-8	1117	8 Wheels
2053	1117	1 Body Cutter

BRIGGS & STRATTON

Part No.	Description
2017-2R	100-2R-6-7 Model 60102, Type 0226-01
2017-3RB	100-3RB-7 Model 80332, Type 0474-01
2017-3RP	800-3RP-8 Model 80202, Type 0301-2
2017-3RB	800-3RB-8 Model 80332, Type 0474-01

KOHLER

2017-4K	4K-7 Model K 91
2017-4K	4FK-7

HONDA

2017-3.5	100-3.5 R-7 3.5 HP Model G150

*Standard Hardware items may be purchased locally

Trouble Shooting Chart

Problem	Cause
1 Engine fails to start.	**A** Check fuel tank for gas.
	B Spark plug lead wire disconnected.
	C Throttle control lever not in the starting position.
	D Check spark plug.*
	E Carburetor improperly adjusted. Engine flooded. Remove spark plug, dry the plug, crank engine with plug removed, and throttle in off position. Replace spark plug and lead wire and resume starting procedures
	F Old/stale gas. Drain and refill with fresh gas.
2 Hard starting or loss of power.	**A** Spark plug wire loose.
	B Carburetor improperly adjusted.*
	C Dirty air cleaner.*
3 Operation erratic.	**A** Dirt in gas tank. Drain, clean and refill.
	B Dirty air cleaner.*
	C Water in fuel supply. Drain and refill.
	D Vent in gas cap and/or carburetor plugged. Clear vent.
	E Carburetor improperly adjusted.*
4 Occasional skip (hesitates) at high speed.	**A** Spark plug fouled, faulty or gap too wide.*
	B Carburetor improperly adjusted.*
	C Dirty air cleaner.*
5 Idles poorly.	**A** Carburetor idle speed too slow.*
	B Spark plug gap too close.*
	C Carburetor idle mixture adjustment improperly set.*
6 Engine overheats.	**A** Adjust carburetor.*
	B Remove any obstructions from air passages in shrouds.
	C Clean cooling fins.
	D Fill crankcase to proper oil level.

Note: For repairs beyond the minor adjustments listed above, please contact your local Lawn Mower Service Shop

For Engine Service see Authorized Briggs & Stratton Service Center

Refer to the Briggs & Stratton Engine Manual enclosed

Questions for Analysis and Discussion

1. How much technical background or knowledge is the reader of this manual assumed to have? How does the organization of the manual help or hinder the inexperienced user?

2. What purposes do the illustrations serve in this pamphlet? How useful do you think they will be? What are the differences between the drawings and the photographs? In particular, what is the value of the exploded diagram in the Lawn Edger and Trimmer Parts Illustration (p. 106)? How do the illustrations in this manual differ from those in "How to Change Oil" (pp. 83–88)? To what extent are these differences a function of the different audiences and purposes of each manual?

3. How does the Parts List included here differ from the specifications for the PHILIPS 2001 (pp. 56–57) in purpose and organization?

4. Using the criteria described in "Here Is What Mechanics Want in Maintenance Manuals" (pp. 123–25), evaluate this pamphlet. Discuss how each aspect of the pamphlet contributes to or interferes with the operator's ease of performance.

Application

Examine the troubleshooting section of this manual. Using an appropriate format, write a troubleshooting guide to help incoming freshmen enroll for classes during your college's registration period.

Taste

To ensure the validity of their results, scientists and laboratory technicians follow standard procedures when performing many tests and analyses. These procedures are fully described in handbooks and reference manuals. One such handbook is *Standard Methods for the Examination of Water and Wastewater,* from which the following section on taste is taken. Like all standard laboratory methods and procedures, these tests for water taste describe precisely the apparatus, procedures, and test conditions to be employed. But unlike other characteristics of water and wastewater, such as pH or biochemical oxygen demand, taste is not easily quantifiable. Thus, these instructions begin with a discussion of the general principles of taste itself and the purposes of taste testing.

211. Taste

1   Taste, like odor, is one of the chemical senses. Most of the general principles of sensory methods described in Section 206 (Odor) apply equally to the taste determination and should be reviewed as background to this section.

2   The differences between the two sense modes are reflected in their corresponding measurement methods. Taste and odor differ in the nature and location of the receptor nerve sites: high in the nasal cavity for odor, and primarily on the tongue for taste.[1] The odor sensation is stimulated by vapors without physical contact with a water sample, while taste requires contact of the taste buds with the water sample. Taste is simpler than odor—there may be only four true taste sensations: sour, sweet, salty, and bitter. Dissolved inorganic salts of copper, iron, manganese, potassium, sodium, and zinc[2] can be detected by taste. The taste sense is moderately sensitive. Concentrations producing taste range from a few tenths to several hundred milligrams per liter. The complex sensation experienced in the mouth during the act of tasting is a combination of taste, odor, temperature, and feel; this combination is often called flavor. Taste tests usually have to deal with this complex combination. If a water sample contains no detectable odor and is presented at near body temperature, the resulting sensation is predominantly true taste.

3   It may not be assumed that a tasteless water is most desirable; it has become almost axiomatic that distilled water is less pleasant to drink than certain high-quality waters. Accordingly, there are two distinct purposes of taste
tests. The first is to measure taste intensity by the so-called threshold test.
The test results are used to assess treatment or pollution abatement required
to convert a water source into a quality drinking water supply or to measure
the taste impact of specific contaminants.[3] The second purpose of taste testing
is to evaluate the consumer's judgment of the quality of a drinking water.
This test involves a panel evaluation of undiluted samples presented as ordi-
narily consumed.[3] A mean acceptance rating of the sample is determined,
based upon a specified rating scale.[4]

4   *Values representing mean thresholds or quality ratings for a laboratory panel are only estimates of these values for the entire consuming population.* [5]

5   Taste tests are performed only on samples known to be safe for ingestion.
Samples that may be contaminated with bacteria, viruses, parasites, or toxic
chemicals such as arsenic dechlorinating agents, or that are derived from an
unesthetic source, are not used for taste tests. A laboratory performing taste
tests must observe all sanitary and esthetic precautions with regard to apparatus and containers contacting the sample. Hospital-level sanitation of these
items and of the small containers for the taste sample must be observed scru-
pulously. Panel taste tests are not performed on wastewaters or similar un-
treated effluents.

6   Use the procedures described in Section 206 with respect to purity of taste
and odor-free water and use of panels of observers.

211 A. Taste Threshold Test

7   *1. **General Discussion***   The threshold test is used when the purpose is
quantitative measurement of detectable taste. When odor is the predominant
sensation, as in the case of chlorophenols, the threshold odor test of Section
206 takes priority.

8   *2. **Apparatus**   a. Preparation of dilutions:* Use the same dilution system as that
described for odor tests in preparing taste samples.

9   *b. For tasting:* Present each dilution and blank to the observer in a clean
50-ml beaker filled to the 30-ml level. An automatic dishwasher supplied
with water at not less than 60 C is convenient for sanitizing these beakers
between tests.

10   *c. Temperature control:* Carefully maintain sample presentation temperature
of 40 C by use of a water bath apparatus.

11   *3. **Procedure***   Prepare a dilution series (including random blanks) as described in Section 206 and bring to the test temperature in the water bath.
Present the series of unknown samples to each judge. Pair each sample with
a known blank sample, both containing 30 ml of water in the 50-ml beaker.
Have the judge taste the sample by taking into the mouth whatever volume
of sample is comfortable, holding it for several seconds, and discharging it

without swallowing the water. Have the judge compare the sample with the blank and record whether a taste or aftertaste is detectable in the unknown sample. Submit the samples in an increasing order of concentration until the judge's taste threshold has been passed.

12   Calculate the individual threshold and the threshold of a panel in the manner described for threshold odor tests.

211 B. Taste Rating Test

13 *1. **General Discussion***   When the purpose of the test is to estimate taste acceptability, follow the taste rating procedure described below. This procedure has been used with water samples from public sources in laboratory research and consumer surveys in order to recommend standards governing mineral content in drinking water.[6] In this procedure, each judge (tester) is presented with a list of nine statements about the water, ranging on a scale from very favorable to very unfavorable. The tester's task is to select the statement that best expresses his opinion. The scored rating is the scale number of the statement selected. The panel rating is the arithmetic mean of the scale numbers of all judges.

14 *2. **Apparatus***   *a. Preparation of samples:* Samples for this test usually represent public finished water ready for human consumption; however, experimentally treated water may be used *if the sanitary requirements given in the introductory material of Section 211 are met fully.* Taste- and odor-free water, and a 2,000-mg/l solution of NaCl prepared with taste- and odor-free water, are recommended as reference samples. ∎

15   *b. For tasting:* Present each sample to the observer in a clean 50-ml beaker filled to the 30-ml level. An automatic dishwasher supplied with water at not less than 60 C is convenient for sanitizing beakers between tests.

16   *c. Temperature control:* Present samples at a temperature that the judges will find pleasant for drinking water; maintain this temperature by a water bath apparatus. A standard temperature of 15 C is recommended, but in any case, do not allow the test temperature to exceed tap water temperatures that are customary at the time of the test. Always specify the test temperature in the test results.

17 *3. **Procedure***   For test efficiency, a single rating session may contain up to 10 samples, including the reference samples noted above. Judges will work alone after receiving thorough instructions and trial or orientation sessions followed by questions and discussion of procedures. Select panel members on the basis of performance in trial sessions. Rating involves the following steps: *a)* initial tasting of about half of the sample by taking the water into the mouth, holding it for several seconds, and discharging it without swallowing; *b)* forming an initial judgment on the rating scale; *c)* a second tasting conducted in the same manner as the first; *d)* a final rating made for the sample and the result recorded on the appropriate data form; *e)* rinsing the mouth with taste- and odor-free water; and *f)* resting 1 min before repeating Steps

a through *e* on the next samples. Independently randomize sample order for each judge. Allow at least 30 min of rest between repeated rating sessions. Judges should not know the composition or source of specific samples. Use the following scale for rating and record ratings as integers ranging from one to nine, with one given the highest quality rating.

18   Calculate the mean and standard deviation of all ratings given each sample.

19   4. ***Rating Scale***   *Action tendency scale:*

1) I would be very happy to accept this water as my everyday drinking water.
2) I would be happy to accept this water as my everyday drinking water.
3) I am sure that I could accept this water as my everyday drinking water.
4) I could accept this water as my everyday drinking water.
5) Maybe I could accept this water as my everyday drinking water.
6) I don't think I could accept this water as my everyday drinking water.
7) I could not accept this water as my everyday drinking water.
8) I could never drink this water.
9) I can't stand this water in my mouth and I could never drink it.

211 C. References

1. **Geldard, F.A.,** 1972. The Human Senses. John Wiley & Sons, New York, N.Y.
2. **Cohen, J.M., L.J. Kamphake, E.K. Harris and R.L. Woodward.** 1960. Taste threshold concentrations of metals in drinking water. *J. Amer. Water Works Ass. 52:660.*
3. **Bruvold, W.H., H.J. Ongerth and R.C. Dillehay.** 1967. Consumer attitudes toward mineral taste in domestic water. *J. Amer. Water Works Ass. 59:547.*
4. **Bruvold, W.H.** 1968. Scales for rating the taste of water. *J. Appl. Psychol. 52:245.*
5. **Bruvold, W.H.** 1970. Laboratory panel estimation of consumer assessments of taste and flavor. *J. Appl. Psychol. 54:326*
6. **Bruvold, W.H., H.J. Ongerth and R.C. Dillehay.** 1969. Consumer assessment of mineral taste in domestic water. *J. Amer. Water Works Ass. 61:575.*

211 D. Bibliography

Cox, G.J. and J.W. Nathaus. 1952. A study of the taste of fluoridated water. *J. Amer. Water Works Ass. 44:940.*

Lockhart, E.E., C.L. Tucker and M.C. Merritt. 1955. The effect of water impurities on the flavor of brewed coffee. *Food Res. 20:598.*

Campbell, C.L., R.K. Dawes, S. Deolalkar and M.C. Merritt. 1958. Effect of certain chemicals in water on the flavor of brewed coffee. *Food Res. 23:575.*

Cohen, J.M. 1963. Taste and odor of ABS in water. *J. Amer. Water Works Ass. 55:587.*

Bruvold, W.H. and R.M. Pangborn. 1966. Rated acceptability of mineral taste in water. *J. Appl. Psychol. 50:22.*

Bruvold, W.H. and W.R. Gaffey. 1969. Rated acceptability of mineral taste in water: II. Combinatorial effects of ions on quality and action tendency ratings. *J. Appl. Psychol. 53:317.*

Bruvold, W.H. and H.J. Ongerth. 1969. Taste quality of mineralized water. *J. Amer. Water Works Ass. 61:170.*

Bryan, P.E., L.N. Kuzminski, F.M. Sawyer and T.H. Feng. 1973. Taste thresholds of halogens in water. *J. Amer. Water Works Ass. 65:363.*

Questions for Analysis and Discussion

1. The concept of taste contains a large psychological component. How have the writers attempted to objectify taste so that it can be quantified? How adequate do you find the definition of taste presented in this article? How does the definition qualify the two tests that follow it?

2. What features of language and organization ensure that if these instructions are followed carefully, the results will be scientifically acceptable?

3. Compare these instructions for performing an experiment with the description of an experiment that researchers have already conducted in the "Materials and Methods" section of "Dwarf Sumac as Winter Bird Food" (p. 202). What are the differences, in language and content, in the ways the two processes are presented? What can you infer about the audience and purpose of each selection that might account for these differences?

4. Compare the language of the Rating Scale (p. 113) of the Taste Rating Test with the language used to discuss the test's apparatus and procedure. What aspects of word choice indicate that the scale was written to be understood by anyone who might be a test subject? What features show that the remainder of the discussion was written for experienced technicians?

Application

Select a specific and straightforward process, such as the installation of a particular piece of equipment or the determination of a certain value. Write a thorough and completely unambiguous set of instructions for the process. Remember, your goal is to ensure that all performers of the process will obtain exactly the same results.

Carcinogens: What the Employer Must Do

In recent years, growing attention has been focused on health hazards in the workplace, and many aspects of worker safety are now protected by federal regulations. These regulations, written in legalistic language, often must be summarized in order to be understood and implemented by the employer. The following guidelines concerning the presence of carcinogens in the workplace are from the Occupational Safety and Health Administration (OSHA) pamphlet, *Carcinogens,* part of the Job Health Hazard Series. This pamphlet condenses regulations appearing in the *Federal Register,* Part III, Volume 39, No. 20, January 29, 1974. Its purpose is "to alert employees to the dangers involved in handling these substances and to remind employers of personal responsibilities for their safe protection and use." The pamphlet explains, in simple language, what the carcinogen is, where it is found, what health hazards it poses, and how it can be controlled.

Establish Regulated Area

1   The employer must set aside a regulated area in which these substances are produced or used. Only specifically trained and equipped personnel shall be allowed to enter the area, and a daily roster of employees entering the area shall be maintained.

2   The rosters, or a summary of them, must be retained for 20 years, and released upon request to the Director of the National Institute for Occupational Safety and Health (NIOSH). If the company goes out of business without a successor, the rosters must be sent to the Director by registered mail.

Air Flow

3   A regulated area must be under negative air pressure—that is, air must flow into the regulated area from any adjacent area and not away from it.

Prohibited Activities

4   Within the regulated area, there must be no eating, storage of food or drink, or storage of containers for food or drink. Storage or use of cosmetics is pro-

hibited. Smoking or chewing tobacco, or storage of tobacco or other materials for chewing, is prohibited. Drinking fountains are prohibited. No sweeping or dry mopping is permitted.

Required Facilities

5 Where washing or showering is required by the regulations, appropriate facilities must be installed. Toilets within the regulated area must be in a separate room. Where employees are required to wear protective clothing or equipment, clean changing rooms must be provided.

Required Signs

6 All entrances to a regulated area must have signs reading:

"Cancer-Suspect Substance
Authorized Personnel Only"

7   When direct contact with these substances could occur (after leaks or spills, or during maintenance or repair of contaminated systems), signs shall be posted stating:

"Cancer-Suspect Substance Exposed in This Area.
Impervious Suit Including Gloves, Boots and
Air-Supplied Hood Required at All Times.
Authorized Personnel Only."

8   In addition, appropriate signs and instructions, in two-inch letters, must be placed at entrances and exits, informing employees of the procedures required on entering or leaving the area.

Required Labels

9 Containers of cancer-suspect substances (carcinogens) must be properly labelled, with lettering on cautionary labels not less than half the size of the largest lettering on the package, but in no case smaller than eight point type. All containers must carry the warning: *Cancer-Suspect Substance.*

10   Containers of carcinogens that are accessible to employees who have not received full training in the nature and danger of exposure must carry the full chemical name and the Chemical Abstracts Registry Number.

11   If containers are accessible only to fully trained employees, the identification may be limited to a generic or proprietary name, or other proprietary identification.

12   Further, there must be no statement on or near either the signs or the labels that contradicts or detracts from the warnings or instructions.

Employee Training

13 Before being allowed to enter a regulated area, each employee must have received specific training in the following areas: the cancer-causing possibilities

of these substances; the kind of assignments that could result in exposure; decontamination procedures; emergency procedures; the employee's specific role in recognizing situations that might result in the release of these substances; and first aid procedures.

14   This training must be repeated yearly following the initial indoctrination. All training materials and related information must be available, on request, to the Director of NIOSH.

Emergencies

15   In an emergency, the regulated area shall be evacuated immediately, and normal operations shall not be resumed until decontamination is accomplished.

16   Employees known to have contacted any of these substances must shower as soon as possible (unless physical injuries prohibit it). Special surveillance by a physician must be initiated within 24 hours for an employee present in a potentially affected emergency area, and an incident report must be filed with the Director of NIOSH.

Physical Examinations

17   Each employee assigned to a regulated area must be enrolled (at no cost to the employee) in a program of medical surveillance, including preappointment examination by a physician.

18   The program must include recording and consideration of a personal medical history of the employee, the employee's family, and the employee's occupational background, as well as significant genetic and environmental factors. The medical examination must be repeated annually.

Medical Records

19   Complete and accurate medical records must be maintained for as long as the employee is employed. When the employee ceases employment, regardless of reason, but including retirement or death, or if the employer goes out of business without a successor, the records must be forwarded by registered mail to the Director of NIOSH. The Director may request such records at any time, as may the employee's personal physician.

Prohibited and Regulated Systems

20   The use of carcinogens in an open-vessel system is prohibited. If these substances are used within a closed system, such as a piping system, with sampling ports or other openings (even if closed when the chemical is in use), employees must wash hands, forearms, faces, and necks before leaving the regulated area or before beginning other tasks.

21   Operations involving a "laboratory type hood," or operations involving transfer of these substances between other than closed systems or containers, require special restrictions, such as continuous local exhaust ventilation, protective clothing, respiratory equipment, and personal hygienic measures. Specific requirements are contained in each standard.

Ventilation

22 Each operation involving these substances must be provided continuous ventilation so that air is always flowing from non-regulated work areas toward the regulated operations. Exhaust air cannot be vented into any area—including the regulated area or the general atmosphere—unless it is decontaminated first, either thermally or chemically.

Protective Clothing

23 Employees must wear clean, full-body protective clothing (smocks, coveralls, or trousers and long-sleeved shirts) as well as shoe covers and gloves before entering a regulated area.

24   Employees handling these substances must be equipped with a half-face, filter respirator effective against dusts, mists, and fumes. On leaving a regulated area, employees must remove protective clothing and gear at the point of exit. When leaving the regulated area for the day, employees must place such clothing and gear in impervious, properly labelled containers for decontamination or disposal.

Washing and Showering

25 Before leaving a regulated area, employees must wash their hands, forearms, faces, and necks. Prior to leaving for the day, they must take showers.

Maintenance and Repair

26 Only specifically trained employees can maintain or repair equipment contaminated with these substances, including leaks or spills. Where such work could result in direct contact with the chemical, employees must wear clean, impervious garments, gloves and boots, and a hood continuously supplied with clean air. Before removing protective gear, employees must be decontaminated. Following removal, employees must shower.

Questions for Analysis and Discussion

1. Given the objectives of this pamphlet, do you think it presents its subject in excessive, sufficient, or insufficient detail? What would be the advantages or disadvantages of presenting more or less detail? If you were an employer, how would you respond to the amount of detail provided in this publication?

2. Why does this subject lend itself to topical rather than chronological organization? Is there any logic to the order in which each topic is presented? Can you group these topics into three or four general areas? Would the pamphlet be better arranged in this way?

3. How do the writers incorporate various techniques of definition into the pamphlet? Do you think all appropriate terms are defined sufficiently? What conclusions can you draw from this selection about the importance of definition in technical writing?

4. Here is an excerpt from the full OSHA regulation for handling carcinogens, as presented in the *Federal Register:*

> *At no cost to the employee, a program of medical surveillance shall be established and implemented for employees considered for assignment to enter regulated areas, and for authorized employees. (1)* Examinations. *(i) Before an employee is assigned to enter a regulated area, a preassignment physical examination by a physician shall be provided. The examination shall include the personal history of the employee, family and occupational background, including genetic and environmental factors.*
>
> *(ii) Authorized employees shall be provided periodic physical examination, not less often than annually, following the preassignment examination.*
>
> *(iii) In all physical examinations, the examining physician shall consider whether there exist conditions of increased risk, including reduced immunological competence, those undergoing treatment with steroids or cytotoxic agents, pregnancy and cigarette smoking.*

This excerpt corresponds to the section on "Physical Examinations" in the *Carcinogens* pamphlet. Discuss the specific ways in which the description of the regulations has been changed to make them easier to understand and implement.

Application

Using topical organization, write a set of guidelines for developing effective study habits, preparing for a job interview, budgeting your money, or another subject that lends itself to such organization.

How to Plan the Entire Product Support Program

A product support program contains material that explains such things as installation, operation, service, and maintenance procedures for a particular product. Examples include IBM's IMS/VS manual, from which the description of data base (pp. 70–76) is excerpted, and the operator's manual for the McLane trimmer and edger (pp. 98–108). The Ken Cook Co. produces such technical support literature for many firms. This short brochure, part of their own product support, was written for executives to explain in general terms how to plan and manage such a program.

■

1   A total product support program consists of an array of product literature and training programs, including:

- promotional literature
- operator and owner manuals
- service and repair manuals
- government documentation
- flat rate manuals
- parts catalogs
- training programs
- price schedules

2   Because product support can represent a significant investment, it is essential that the product support program be carefully planned and managed for maximum efficiency and effectiveness.

What Are the Advantages of a Total Program?

3   - Multiple use of the same research, write-ups and illustrations for promotion, operation, service and training.
- Standardization of terminology and information.

- Coordination of projects to ensure timely completion of all components of product support.
- Projection of a common image to dealers and customers.
- Elimination of duplicate efforts—resulting in lower cost.

How Can You Manage a Total Program?

4 - Establish a task force composed of personnel from different functional departments or groups. This task force should evaluate product support requirements and establish priorities.
- Select an individual as leader of the task force to maintain ultimate management responsibility over the program. This individual will be responsible for scheduling, manpower allocation, and budget control.
- Assign an editorial or coordinating staff to ensure standardization of information and prevent duplication of effort.

How Can You Schedule a Total Program?

5 - Begin by examining your needs and goals.
- Establish dates by which you *must* have final product support components. These dates usually depend on some other event—release of a product for sale, a trade show, a regularly scheduled dealer training seminar, etc.

How Can You Staff for a Total Program?

6 - Establish realistic dates for beginning each product support program. Ideally, effort should not begin until the product prototype is finalized and ready for production and all reference materials (including engineering drawings) are complete. Of course, this is not always possible.
- Mark the realistic start-of-work date and final due date for each product support component on a planning chart or calendar. Make sure enough time is available to complete each project using available facilities and personnel. If not, some adjustments must be made.
- For the fortunate, efforts for a product support program will last an entire year and will require an even distribution of man-hours throughout the year. For the majority, however, the product support efforts will be concentrated in several months, moderate in other months, and very slow in others.

7 Maintaining the right size staff for a fluctuating workload is not easy. Here are some suggestions for dealing with the manpower problem:

8 - Subcontract all or portions of a project to a qualified technical publishing firm.
- Ask employees to contribute overtime hours to meet scheduling requirements.

- Employ part-time help during peak load period. Contact a technical services job shop, or consult a college placement office to locate students interested in part-time work.

Conclusion

9 Treat product support as a total, planned program, rather than as isolated components which must be provided. Properly managed, a total program will contribute to your overall efficiency and profitability.

Questions for Analysis and Discussion

1. What aspects of the language, sentence structure, and content of this brochure indicate that it was written for an executive audience?

2. Is the use of questions an effective tool in invoking the reader's interest in this pamphlet? Do these questions help organize the information?

3. Compare this as a topically ordered set of instructions with the instructions on keeping the workplace safe in "Carcinogens." Contrast the levels of detail and generality in each, and discuss your conclusions in light of the purpose of each document.

Application

On the basis of this example, what things would you consider in producing a brochure? Design a brochure aimed at attracting incoming freshmen to your major field of study.

Clyde Cheney

Here Is What Mechanics Want in Maintenance Manuals

This article on how to write a maintenance manual was published as an "open letter to manual writers" in *Plant Engineering,* a technical journal. Not only does the article summarize in topical form those qualities that make manuals especially helpful for maintenance mechanics, it is itself a good example of a logically organized, clearly explained set of instructions.

1   If all maintenance mechanics in the industrial world had the opportunity to voice their opinions to manual writers, they probably would make statements like these: "Please, won't somebody listen to us. We'd like to do the best job we can, but it is difficult to understand the manuals we usually get with machines and components.

2   "When something breaks down, we have to fix it in a hurry. We do fairly well most of the time, but nobody knows everything, and we have to look for help now and then. It's really nice when there is a *good* manual around, especially on the late shift."

3   **Here is what mechanics would like to have in a manual:**

4   - Organize the information so it can be found in seconds. Use tabs for sections on the different components of a machine, like the feeder, drive, takeoff, controls, ventilation, cooling, etc.

5   - Set up a table of contents or index with page numbers. Chapter and section numbers or letters do not tell enough; page numbers do. And put the page numbers where they can be seen.

6   - Use pictures, sketches, drawings, or any other illustration to help keep down the number of words to describe what can't be pictured easily, Fig. 1.

7   - Use plain language in the text.

8   - Put illustrations near the text or explanation. If mechanics have to leaf back and forth to find a figure number, forget it: you are wasting your time and theirs.

9   – Remember that text in narrow columns is much easier to scan than text in full page width.

10   – Use large print so mechanics can read it in dim areas of a plant.

11   – Include a good troubleshooting chart when possible. Put it in the front of the manual where it can be found quickly. And put a tab on it. After all, the main reason for picking up a manual is to find helpful answers to a problem without any undue delay.

12   – Use exploded views when the text relates to replacement of parts, subassemblies, etc. They prove very helpful when mechanics have to tear down a component to fix it. Exploded views best show exactly which piece goes where, especially if the mechanic has to put together something that somebody else took apart—and that person is not available to explain what he did. Fig. 2.

13   – Print on both sides of each sheet. It also helps if the manual is made up of loose-leaf sheets so the whole page is visible and each page can be replaced easily by revisions. The paper should be heavy enough so the sheets don't tear easily.

14   – Use a different color to highlight a warning or cautionary note to call the mechanic's attention to safeguards that prevent damage to equipment or harm to the mechanic, Fig. 3, It is always good to know, for example, that a screw is springloaded and can go flying if it is backed off too far.

15   – Take time to underline or somehow identify by number or description the model that the manual covers—please. Sometimes manuals cover more than one model and that can be confusing. It isn't always easy to find a data plate on the machine. And the mechanic may not have a material list with the correct identity.

16   **Things mechanics do not like to see in manuals:**

17   – Copies of photos or blueprints. A poor picture or print cannot be made better on any copier.

18   – Extensive use of decimal numbers for page, chapter, or paragraph identification, Fig. 4. A simple outline is much better than endless decimal number references.

19   – The placement of sketches and drawings vertically on the page, requiring the rotation of the book or person to read it. It is difficult to follow such a diagram or drawing while both hands are employed with the tools or component being serviced. Everything on a page should be placed so it can be read from one direction.

20   – Long-winded sentences and big words. The manual does not need a lot of theory; all explanations should be kept simple and to the point.

21   – Cluttered illustrations. A full page should not be used when only a small portion is needed to get the point across.

22   These are a few examples of what mechanics do and do not need in manuals. They'd like to do their jobs well; manufacturers and manual writers can help by doing their best to get *good* manuals to the maintenance mechanics.

Figure 1 A picture or sketch of a component (motor brake assembly) can save a lot of unnecessary words and prove to be more helpful.

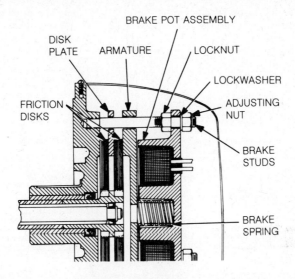

Figure 2 Exploded views best describe where each part belongs in an assembly. The parts are easily identified from the parts listing.

BOTTOM BLOCK ASSEMBLY

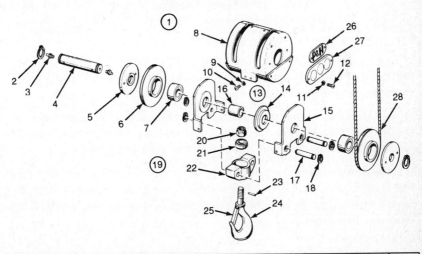

Ref. No.	Description	Part Number	Qty.
1	Bottom Block Assembly—w/o Safety Latch .... Includes Ref. 2 thru 15	3100E13931-13	1
	Bottom Block Assembly—w/Safety Latch .... Includes Ref. 2 thru 16	3100E1393-4	1

Figure 3 Whenever it is necessary to include a "warning" or "caution" notion in the manual, a special color calls it to the mechanic's attention.

```
┌─────────────────────────────────────┐
│          ┌──────────┐                │
│          │ WARNING  │                │
│          └──────────┘                │
│  Before starting any disassembly of the hoist, lower the
│  bottom block to the floor, if possible. If the bottom
│  block cannot be lowered to the floor, secure it to the
│  beam or other mounting support to relieve the tension
│  on the wire rope. If the hoist is being removed from its
│  mounting, shut off the power supply, and disconnect
│  the power connections to the hoist. If the hoist is to be
│  worked on while still mounted in place, shut off the
│  power and pull the power supply fuses or place the
│  power supply circuit breakers in the OFF position.
│
│          ┌──────────┐                │
│          │ CAUTION  │                │
│          └──────────┘                │
│  Do not immerse prelubricated bearings in cleaning sol-
│  vent. Never allow a bearing to spin when drying it with
│  compressed air.
└─────────────────────────────────────┘
```

Figure 4 After reading the many references to different decimal numbers, many mechanics shake their heads in disgust. A simple outline would be better.

38.1.4.3 If these two readings do not result in an axial float of 0.003 in. to 0.005 in., substitute a different thickness of No. 32 shims and repeat paragraphs 38.1.3 with new thickness shim, also 38.1.4.1 and 38.1.4.2, until the reading falls in this range. Tap end of bevel pinion with plastic hammer to seat bearing No. 78 after shimming.

38.1.5 With an inside micrometer or equivalent, measure "A" while applying hand force on high-speed shaft so as to pull it away from center of housing (taking up bearing axial float). List dimension.

38.1.6 Add dimension "A" to 2.6876 in. and compare this with the mounting distance etched on the bevel pinion or gear. The mounting distance should be equal to etched figure or 0.002 in

Questions for Analysis and Discussion

1. What are the qualities that Cheney suggests make a good maintenance manual? As a technical writer, how helpful do you find this article? What features make it useful to you?

2. What qualities of format, style, and content make this article a good example of instructional writing? To what extent does the writer follow his own advice?

3. What elements of the language give the article its conversational tone? How does Cheney give the impression that he is speaking directly to the reader? How does this make the text more readable?

4. Cheney emphasizes the importance of format, graphics, and the physical qualities of the manual itself. Indeed, he gives more attention to these topics than to writing and language. What are the practical reasons for this emphasis?

Application

Write and design a short pamphlet that offers complete instructions for a process with which you are familiar. Use all you know about headings, graphics, white space, and layout to make your instructions clear and easy to follow.

4

PROPOSALS

Proposals and grant applications explain, in detail, how a project will be done, how much it will cost, and why the individual, firm, or institution proposing to undertake the project should be selected to do it. In other words, a proposal is a bid for a contract to do something: to study a problem, develop a product, acquire a piece of equipment, implement a program, conduct original research, or simply accomplish a specific task. There are as many types of proposals as there is work in the world to be done. Proposals are the vehicles that link those who wish to do the work with those who want it done or believe it should be done.

Proposals may be solicited or unsolicited. Solicited proposals are common in municipal, state, and especially federal government, as well as in private industry. For instance, the SCS proposal on pp. 135–40 was prepared for Fresno County in California. A solicited proposal may be submitted in response to a "Request for Proposal," commonly referred to as an RFP, issued by the agency or business wishing to have work performed. A general RFP may be published, allowing open competition among all qualified firms or institutions that wish to submit proposals, or a single firm or group of firms may be asked to describe how they would do the job.

RFPs may be issued in conjunction with a "Statement of Work" (SOW), which describes what the proposed contract will involve. The company most likely to win the contract is that company which has best understood and responded to the SOW.

Increasingly common is a "Request for Qualifications," or RFQ, preceding the issuance of the RFP. In this case, the potential contractor asks each competitor to submit a qualifications package. This package, similar to the management and capabilities section of a full-form proposal, includes information about the company: its history, philosophy, organization, personnel, facilities, and experience. After the qualifications packages have been reviewed, one or a number of companies will be asked to submit complete proposals.

An unsolicited or grant proposal is submitted to likely sources of financial assistance in response to a need or problem perceived by the proposer. Many government agencies, such as the National Science Foundation (NSF) or the National Endowment for the Humanities (NEH), as well as other public and private institutions, fund these proposals. The grant proposal for research on two endangered species of butterflies, on pp. 160–63, is such a proposal.

Proposals for advanced research are commonly written in research facilities and universities. But many community organizations or groups also write unsolicited proposals to obtain goods and services. The goals of the unsolicited or grant proposal are more complex than those of the solicited proposal. The writer must first convince the reader that the proposed project should be done. Then, as in solicited proposals, the writer must also convince the reader that he or she is capable and qualified to complete the project satisfactorily.

Most proposals are formal documents organized according to fairly uniform standards of content and organization. Often, however, proposals may be submitted as a letter or a memorandum, as in the case of "add-on" proposals—statements of proposed extensions of on-going projects. Short proposals written within a company or organization and requesting authorization to purchase equipment, as in the Shaw Hygrometer proposal (p. 143), or to implement new procedures, as in the recommendation for a floor-topping project (p. 146), are frequently written as memoranda. Short proposals written outside the writer's company or organization often are in the form of letters, as in the RERC proposal to conduct a marketing analysis (pp. 149–50).

Full-form proposals, regardless of their length, usually consist of a technical proposal and a cost proposal. The technical proposal discusses the problem and its solution: what is to be done and how it is to be accomplished. The cost proposal, or budget, presents an itemized account of project costs.

The cost proposal is fairly straightforward. It may be a separate proposal or section or may simply consist of a form supplied by the contracting organization. In shorter proposals, as in the SCS example (pp. 135–40), the cost proposal may be incorporated into the final pages of the text itself. The main consideration in preparing a cost proposal is to estimate accurately the costs in personnel and materials (including fees paid to consultants, if needed). The more difficult task of persuading the contractor to accept the bid is reserved for the technical proposal.

The technical proposal details the substance of the project itself. Figure 4.1 (pp. 129–31) shows a table of contents for a typical full-form technical proposal. The technical proposal has two distinct parts: A careful and detailed consideration of the problem and its proposed solution, and a presentation of the management and capabilities of the organization proposing to undertake the work. In the first section, the writer aims to convince the contracting agency that his or her organization has a creative grasp of the problem, the technical competence to solve it, and a well-thought-out plan of operation. All claims made in this section of the proposal must be supported by concrete data. The following may be included in the first section:

TECHNICAL PROPOSAL

DEVELOP A MANUAL FOR UPGRADING
EXISTING UTILITY DISPOSAL
FACILITIES IN RESPONSE TO
REQUEST FOR PROPOSAL RFP 1685-2

Issued by the Fossil Fuel and
Advanced Systems Division of
the Electric Power Research Institute

Submitted by

SCS Engineers
4014 Long Beach Boulevard
Long Beach, California 90807

October 26, 1979

CONTENTS

CONTENTS (continued)

Summary

This short but complete summary of the problem and proposed solution is read by more individuals than is any other section. Therefore, it should be written to convince the reader that the proposed solution is the best one and that the writer is fully qualified to complete the work successfully and most economically. In shorter proposals, the summary may be excluded and its substance included in the letter of transmittal, as in the SCS example (pp. 135–36).

Introduction

The introduction provides the reader with basic information concerning the company, its philosophy or policies, its experience, and its general approach to the problem. The introduction also outlines the organizational plan for the rest of the document. The purpose of the introduction is to give the reader everything needed to understand the pages that follow. Both the summary and introduction are similar to their counterparts in the technical report (see pp. 166–67), with this important difference: All material must be designed to persuade as well as inform the reader.

Problem

The background to the problem, the scope of intended work, and a general overview of the writer's treatment of the problem should be included in this section of the proposal. In the solicited proposal, the writer here demonstrates his or her full understanding of the contractor's needs and the contract's specifications. A simple repetition of the RFP or SOW is inadequate. Instead, the proposer should demonstrate a creative and intimate understanding of the client's problem. In the case of an unsolicited proposal or grant application, the background to the problem is explained, together with a demonstration of the timeliness and significance of the issue. In this case, the writer must demonstrate that the problem needs to be addressed, and then must convince the contracting agency or institution to provide funds.

Approach

This section of the proposal details the methods and step-by-step procedure to be followed. The project team may also be introduced, and a task plan and completion schedule presented. Graphics are frequently included to allow the reader to see at a glance that the work plan and time schedule are well-ordered and reasonable. Periodic reports may be scheduled to assure that the contractor will be kept informed of all aspects of the project, and will be able to monitor progress and intervene should unexpected problems or opportunities arise.

The remainder of the technical proposal is devoted to a description of the management and capabilities of the individual, company, or team proposing to perform the project. This part of a proposal is essentially a qualifications package and, with certain modifications, would be submitted in response to an RFQ. Topics to be covered include:

Company and Project Management

This section describes the organization of the project team, the way in which the team fits into the overall management structure of the company, and the lines of responsibility and authority for the project. Frequently, a chart indicating both company and project team organization gives the reader a quick and adequate understanding of the management system.

Personnel

The members of the project team are described in some detail in this section of the proposal. An up-to-date descriptive résumé will usually be maintained for each person likely to be involved in company projects. These résumés will be adapted to suit the requirements of individual proposals. The résumé describes each individual's education, company position, former employment (if related to the project at hand), experience on similar projects, special qualifications for the project, and professional affiliations and publications.

Related Experience

This section describes work performed by the writer that is related to the proposed project and that indicates ability. An efficient way to organize the section is to provide a list of completed or on-going projects, together with the names of the clients for whom the work was or is being done.

Facilities

Those facilities of the company required for the successful completion of the proposed work are described in detail.

Regardless of the subject, content, length, or format, all proposals are essentially sales documents. The goal of a proposal writer is to persuade. And the writer must persuade not through the use of hyperbolic promises—sales "hype"—that can never realistically be kept, but through positive, active assertion of facts supported by examples and experience.

David E. Ross
SCS Engineers

Proposal for Solid Waste Management Planning

The following proposal to determine solid waste generation and composition factors was submitted to the Solid Waste Coordinator for Fresno County, California, by SCS Engineers, a private consulting firm specializing in environmental and waste engineering. As explained in the letter of transmittal, it is the first of three related proposals, which together make up a comprehensive project. A short proposal of limited scope, it describes a study to provide data preliminary to the design and development of a solid waste management program for the county. Although brief, the proposal contains most of the elements of a full-form proposal. Note its positive tone and language: This proposal communicates a picture of a competent and experienced company fully able to complete the proposed work successfully. ■

April 19, 1979
File No. L7943-A, B, and C

Mr. Richard V. Anthony, Solid Waste Coordinator
County Administration Office
Hall of Records, Room 300
Fresno County
2281 Tulare Street
Fresno, California 93721

Subject: Proposals for Solid Waste Management Planning in Fresno County

Dear Mr. Anthony:

1   SCS Engineers is pleased to submit three copies each of three separate outline proposals to support Fresno County's solid waste management planning efforts. The three proposals concern:

 — Solid waste generation and composition study
 — Identification of markets for materials and energy recoverable from Fresno County solid waste
 — Investigation of alternative sites for a new Fresno sanitary landfill.

2   The proposals outline basic project plans. Specific project goals, details of how the projects are to proceed, and refined project scheduling will be defined through discussions with Fresno County officials during contract negotiations. The estimated project manpower loadings and budgets reflect our present perception of the work to be done. These budgets are subject to modification during negotiations, depending on the final work plans developed.

3   Note that all three projects would be performed concurrently. SCS has available the experienced personnel to effectively complete these efforts in this manner.

4   The extensive qualifications of SCS Engineers in solid waste management have been provided to and discussed with you previously. Our capabilities and experience are not included here in the interest of brevity.

5   Fresno County is approaching the complex problem of solid waste management rationally. We would be pleased to provide our assistance to this worthy effort.

6 Please contact the undersigned if you have any questions.

Very truly yours,

David E. Ross
Associate
SCS ENGINEERS

DER: lt

Enclosures

Proposal to Determine

■

WASTE COMPOSITION AND GENERATION IN THE FRESNO COUNTY METROPOLITAN AREA

Presented to

County of Fresno
County Administration Office
2281 Tulare Street
Fresno, California 93721

By

SCS Engineers
4014 Long Beach Boulevard
Long Beach, California 90807

(213) 426-9544

April 20, 1979

Technical Approach

1 Characterization of the Fresno County Metropolitan Area (FCMA) Solid Waste Shed is an integral part of future resource recovery planning efforts. Solid waste characterization as it is used here refers to quantification of the following variables:

- Solid waste generation and flow to existing transfer and disposal sites
- Solid waste physical composition as received (glass, metals, newsprint, etc.)
- Solid waste fuel fraction analysis (heating value, ash and moisture content, proximate/ultimate analysis, etc.)
- Seasonality and projections of solid waste generation.

2   The depth of investigation for each variable will vary with the stages of program implementation and the type of system envisioned. Each variable should nonetheless be quantified in a manner that reflects local solid waste characteristics. Figure 1 displays a conceptual solid waste characterization procedure aimed at resource recovery planning.

3   SCS Engineers proposes to implement just such a comprehensive program for FCMA, suitable for use in the preparation of a feasibility study and conceptual system design. By virtue of an extensive knowledge of available waste characterization procedures, we can provide FCMA with the best data base obtainable for the given level of effort. The following task outline summarizes our proposed approach to waste characterization in the FCMA.

Task A—Waste Characterization Survey Planning

4 SCS will begin the study with a detailed review and description of the FCMA waste shed. This survey will include all existing and proposed disposal sites and collection jurisdictions. We will then proceed to identify waste generation centroids according to population and industrial centers. Both the waste sheds and centroids will be plotted on FCMA maps to assist in locating primary sources of recyclables, as well as optimal transfer station and resource recovery sites. SCS will then confer with Fresno County personnel to determine:

- What alternative systems are under consideration (i.e., what levels of data accuracy and precision are required)
- What waste-related variables, strata, and descriptive units are desired, and
- What data are already available.

Based upon this information, we will prepare a waste characterization program that best suits the needs of FCMA.

5   For waste quantity estimation, any existing data will be researched and assessed for accuracy and completeness. Weight surveys will be planned for all disposal sites that do not currently weigh their solid waste. Historical vol-

ume data will be obtained from all sites for projections and seasonality estimation, regardless of the quantity measurement procedure.

6   Available waste composition data will be obtained from previous studies and FCMA records. These data will be assessed for accuracy and precision, and supplementary field composition surveys will be proposed. Two two-week surveys are anticipated, each survey covering all FCMA landfills within the specified period. The six-month spread should demonstrate differences in solid waste composition between summer and winter. The waste composition surveys may include sorting in the following categories:

- Corrugated cardboard
- Newsprint
- Kraft bags
- White high-grade paper
- Mixed high-grade paper
- Mixed glass
- Ferrous metal
- Aluminum
- Other nonferrous metals
- Plastic, rubber
- Wet garbage
- Fines.

7   The waste sorting procedure will be based upon the technique developed by SCS for EPA and ASTM in 1978. Using that technique, a crew of two SCS supervisors and six laborers will sort an estimated 12 300-lb samples per day for the duration of the survey. For budgeting purposes, it is assumed that CETA workers can be obtained at no cost to the project.

8   Chemical waste characterization is not within the scope of this study. SCS will use the waste composition estimates to generate approximate heating value estimates for both the as-received solid waste and the combustible fraction.

9   Once a plan of action has been selected for each survey aspect, the plan will be reviewed with FCMA and the landfill operators to obtain their approval and cooperation.

Task B—Waste Survey Execution

10   Following the approval of the survey plan, SCS will proceed with the field survey. Waste shed characterization will begin immediately following contract award. A supplementary report describing the waste shed findings will be submitted to FCMA a month later for review and confirmation.

11   The waste quantity surveys will also be implemented as appropriate within the first month after contract award. They will be conducted over a combined total of two weeks per season (summer and winter). Portable platform scales will be used where necessary. Quantity estimation will be based on measured average density and combined truck volumes per day and week. An estimated 75 to 100 percent of all trucks can be weighed at each survey

site using the proposed portable scales, thereby providing an estimate with
± 5 percent precision or better.

12 The waste composition surveys will be conducted on the two weeks following the weight survey. An estimated 120 samples will be collected and
analyzed during each sampling period. Using the waste component categories
mentioned earlier, a final sorting strategy will be selected by SCS and FCMA
within the first month. The data output will be presented both according to
disposal site and FCMA as a whole.

Task C—Analysis of Results

13 The solid waste characterization data obtained during Task B will be compiled immediately after each survey and presented to FCMA in summary
form. In addition, the historical data and survey results will be combined to
provide FCMA with 10-year waste generation projections. The waste shed
data will also be used to provide estimates of waste generation and recyclables
concentrations. These same concentrations, or centroids, will be presented in
a format suitable for use in transfer station siting and waste processing or
buy-back center location.

Task D—Reporting of Results

14 Because activities on a project of this type are cyclic, monthly program submittals are not appropriate. SCS instead proposes the following reports to be
submitted to FCMA over the 8-month project:

- Waste shed characterization report
- First waste survey plan and schedule
- Summary of findings during first waste survey
- Second waste survey plan and schedule
- Summary of second survey findings
- Draft final report
- Final report.

Project Schedule Waste Composition Project

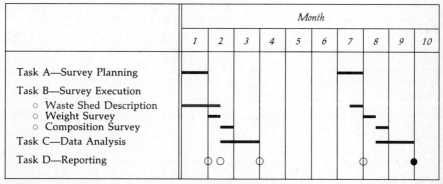

Legend
- Principal Activities   ○ Interim or Preliminary Report
- ● Final Report   ○ Planning Reports

This schedule of submittals, coupled with other routine phone and written communication, will serve to inform FCMA of all project activities in a timely manner.

Table 1   Manpower Estimate Solid Waste Characterization (in person-hours)

Personnel	Task				Total
	A	B	C	D	
Project Director/Manager (J. P. Woodyard)	16	40	24	40	120
Staff Engineer (T. Boston)	40	120	24	16	200
Statistician (H. Rishel)	8	16	24	8	56
Staff Engineer (M. Bulot)	40	120	—	16	176
Technician	—	240	—	—	240
Secretarial	8	24	16	40	88
Graphics	4	16	12	16	48
Totals	116	576	100	136	928

Table 2   Estimated Budget Solid Waste Characterization

Personnel Services	Hours	Rate $/hr	Total
Project Director/Manager	120	$ 36	$4,320
J. Staff Engineer	376	17	6,392
Statistician	56	25	1,400
Technician	240	17	4,080
Secretarial	88	14	1,232
Graphics	48	14	672
Subtotal	928		$18,096

Other Direct Costs

Travel		
Company truck, 2,500 mi @ $0.20/mi	$ 500	
Per diem, 66 days @ $25/day	1,650	
Reproduction—1,000 pp @ $0.07	70	
Telephone (long distance)	200	
Engineering supplies (scale, etc.)	1,400	
Subtotal		$ 3,820
Total		$21,916

Questions for Analysis and Discussion

1. What expectations does the letter of transmittal create in the reader concerning the proposal that follows? How does the letter of transmittal contribute to the proposal's persuasive power?

2. The tasks described in the body of the proposal are divided into self-contained units, each with specific and identifiable objectives. Does this division provide a logical approach to the problem? How is the reader served by this methodical plan of work?

3. How have the writers of this proposal used lists to increase readability and to ensure that important information is clearly communicated?

4. What characteristics of the proposal suggest that SCS will be able to complete the work competently and efficiently? Is this a convincing proposal?

Application

Examine the effectiveness of the project schedule (p. 140) in offering an overview of a well-ordered plan of work. How does this schedule conform to the breakdown of tasks presented in the body of the proposal? Select a project on which you will be working for the next few weeks or months and divide it into major tasks and sub-tasks. Construct a similar chart showing your work schedule. You may want to change the time periods from months to weeks and to add specific dates. How does this schedule make your project seem more or less feasible?

John R. Van Surdam

Memorandum:
Shaw Hygrometer

This memorandum recommends the purchase of a particular piece of equipment. Although it is brief and informal, it contains most of the elements of a proposal. The proposal begins by establishing the problem at hand. Because the writer is addressing another individual within the company, he does not provide much background information. Instead, he centers on justifying his recommendation for the specific purchase proposed. Informal proposals such as this are common in business and industry.

—MEMORANDUM—

TO: Dick Ryburn

FROM: John R. Van Surdam

DATE: July 22, 1980

SUBJECT: Shaw Hygrometer for Monitoring Cargocaire Moistures

1   The irregularities recently encountered in physical wet and dry bulb measurements on the Cargocaire air dryers have caused process moisture control to become a major problem. One solution is to purchase a hygrometer to standardize the moisture reading. A hygrometer will allow us to make control decisions much more quickly and with greater confidence.

2   The hygrometer best suited for our system is the Shaw Model SH4, 4 point manual meter with alarm delay. I recommend purchase of the Shaw Hygrometer for the following reasons:

- Sensors would be located in each of the three 16-inch diameter process air risers from the Cargocaire to determine individual machine performance.
- One sensor would be located in the 20-inch diameter main header to determine overall moisture content.
- Our maximum moisture content of 10 grains/lb. corresponds to a dew point of $-11.11°C$, and would require the medium range ($-60°$ to $0°C$) "Yellow Spot" sensors.
- Accuracy should be ±2 grains, and minimum sensor service life should be 3 to 5 years.

3   The prices quoted below are for this date. Delivery time is 10 weeks.

Model SH4 hygrometer	$5,185
Alarm option	430
Sampling cells	300
Cable	200
Total	$6,115

RVS/jhs

cc: Mr. Ted Folsom
    Mr. Jim Rodgers

Questions for Analysis and Discussion

1. Identify as many features of a proposal as you can in this memo. Which elements sometimes found in a proposal are not present here? How is the presence or absence of these elements a function of the memo's audience and purpose?

2. What particular words or phrases suggest that the writer is confident and emphatic about the merit of his recommendation? Why is this a good tone to adopt in writing a proposal?

3. How does the writer arrange his material on the page in order to help the reader follow the organization of his proposal?

Application

Using memo format, write a brief proposal to your technical writing teacher, suggesting a specific change in the content, schedule, or procedures of the class. Consider your audience and purpose carefully in writing this memo, and state your recommendation in terms that are as positive as you can make them.

Floor Topping Project for X Plant

The following recommendation report proposes awarding a contract for floor resurfacing to "A" company. Like the preceding proposal recommending the purchase of a Shaw hygrometer, this short memo is addressed to readers familiar with the problem concerned. Thus, it can concentrate on explaining logically and thoroughly why "A" company is more suitable for the job than "B" company. The writer of this memo must also justify accepting the higher of two bids. To do this, the memo clearly enumerates the advantages of "A" company and the disadvantages of "B" company. It also shows how contract bids are evaluated on the basis of their response to the contractor's specifications or statement of work.

■

February 15, 1982

File: FLO

John Doe
Purchasing

FLOOR TOPPING PROJECT FOR X PLANT—INQUIRY NO. 1234

Recommendation

1 The 15,000 square feet of epoxy floor resurfacing should be contracted with "A" Company, based on conformance with the specification sheet and anticipated better performance compared to the quote from "B" Company.

Discussion

2 "B" Company quoted $34,600, but did not follow Specification Sheet No. 4321 in the following areas:

1. Saw cuts for perimeter termination to be ¼" × ¼" deep versus the specified ½" wide × ½" deep. This change might cause poor bonding between the epoxy and the concrete at the perimeter.
2. "B" will use flat hand troweling versus the four-bladed power trowels which were specified to compact and level the epoxy overlay. This change leads to air voids in the epoxy and a poor surface finish.
3. Compressive strength of the "B" floor topping will be 13,000 psi versus the specified 16,000 psi. Poor strength of the "B" floor topping will lead to loss of aggregate and poor wear resistance.

3 Gouging tests were made to compare the "B" floor topping to the "A." Compared to the "B" topping, the "A" floor topping has superior resistance to gouging and floor damage. The greater strength of the "A" floor is probably due to graded aggregate and a better epoxy resin.

4 "A" will use the "Blank" concrete cleaning system, which uses steel shot and a dust removal system. This should reduce the dust compared to the normal mechanical scarification.

5 "A" Company quoted $40,000 to resurface 15,000 sq. ft. of floor. Better results are anticipated compared to the quote of $34,600 from the "B" Company.

6 "A" will provide a two-year guarantee on the floor resurfacing which exceeds the specified one-year guarantee on Specification No. 4321.

Jane Smith
Engineering

fg

Questions for Analysis and Discussion

1. This memo is written from Engineering to Purchasing. How has the writer considered the reader's background and needs when presenting technical information?

2. How does the writer organize the Discussion section to support her Recommendation? In particular, why does she choose to begin with the disadvantages of "B" company and end with the advantages of "A" company? What would be the effect of reversing the order of the Discussion?

3. The writer of this memo must be careful to justify her recommendation to accept a higher bid. How are her use of detail and the precision with which she describes both companies effective in supporting her recommendation? Why would her recommendation have been weaker had she simply stated "I believe 'A' company will do a better job than 'B' company"?

4. Why has the writer listed most of her material, with one fact per statement? Is this an effective format for this kind of proposal? Why or why not?

5. This memo summarizes the Badische Corporation's evaluation of two bids to perform work. What does it suggest are the standards used to evaluate bids or proposals? How closely should a bid follow project specifications or the statement of work (SOW)?

Application

Select two journals important in your field of study. Assume your department is able to subscribe to only one of these journals and will make copies available to students in a departmental lounge. Locate copies of both journals in the library and examine them carefully. Write a brief proposal recommending the journal to which you think your department should subscribe. Be careful to establish clearly the criteria you use in making your selection.

RERC Proposal

The following letter from Real Estate Research Corporation (RERC) proposes to conduct a market feasibility analysis and appraisal for a proposed neighborhood shopping center. Unlike the preceding two memos, this letter is not written to someone in the writer's company, but instead is addressed to an outside client. Thus, it must devote more attention to contractual matters, clearly spelling out the terms of the proposed project as well as the rights and obligations of both RERC and the client. Although written as a letter, this selection contains many elements also found in formal proposals.

March 16, 1982

Mr. Samuel Sewell
President
Community Development Corporation
2500 Leavenworth Road
Lansing, Kansas 66601

Dear Mr. Sewell:

1 In accordance with our telephone conversation on March 15, 1982, Real Estate
   Research Corporation (RERC) is prepared to conduct a market feasibility
   analysis for a proposed 50,000-square-foot neighborhood shopping center to
   be built in Lansing, Kansas.

2 Our analysis of the market for the proposed retail center will include a delin-
   eation of the trade area, an analysis of disposable household income and other
   demographic characteristics within the trade area, and an evaluation of competitive properties--both existing and planned. We will provide a list of the
   types of stores that should be included in the center and estimates of achiev-
   able rents and absorption time. We will also prepare financial pro formas
   based on projected income and expenses for the first and third full years of
   operation.

3 We are prepared to conduct this market feasibility analysis for a fee of
   $12,000 plus direct costs. Direct costs include transporation, lodging, per
   diem, telecommunications, and report production. We will provide five copies
   of a fully documented report within six weeks of receiving your authorization
   to proceed.

4 We typically require a retainer of 50% of the professional fee prior to starting
   an assignment. We are prepared to commence work upon the receipt of one
   copy of this letter, signed by you or an authorized representative, together
   with a check for $6,000. This proposal is subject to the attached RERC Stan-
   dard Proposal Terms and Conditions and will remain in effect for 30 days.

5 If our initial field work indicates that our conclusions on the market for the
   proposed center are likely to be negative, we will notify you immediately.
   Should you then decide to terminate the study, you would be billed only for
   the time expended and expenses incurred.

6 Thank you for your confidence in RERC. We look forward to working with
   you on this assignment.

Respectfully submitted,

REAL ESTATE RESEARCH CORPORATION

George Johnson
Vice President

RERC

Standard Proposal Terms and Conditions

7 This section sets forth terms and conditions applicable to the attached proposal. Acceptance of the proposal also constitutes acceptance of these terms and conditions.

Payment Terms

8 The full amount of the fee and direct expenses, less any prior payment, will be due upon completion of the work program and submission of the final report. Payment for services will be due upon receipt of invoice. If payment is not received within thirty (30) days of the billing date, RERC reserves the right to apply a service charge of 1.5 percent per month or fraction thereof to the total unpaid sum. It is further agreed that, in the event legal action becomes necessary to enforce collection of bills, the client will be responsible for all collection costs, including but not limited to court costs and reasonable legal fees.

Confidentiality of Assignment

9 RERC will respect the confidential nature of this assignment and in so doing will use its discretion where specific identification of the project or the client organization might be involved in obtaining research data. Reports are the property of the client and will not be made available to any other organization or individual without the consent of the client. This agreement, however, will not limit RERC from performing assignments of a similar nature for other clients in any area, now or in the future. In any event, all work performed under this assignment would continue to receive confidential treatment.

Use of Reports by Clients

10 RERC's interim drafts, memorandums, and final reports will not be presented to third parties by the client except in the form delivered to the client. No abridgement, abstracting, or excerpting of reports will be made without obtaining the permission of RERC. The copyright to all reports is held by RERC.

Objectivity

11 To protect you and other clients, and to assure that RERC's research results and appraisal values will continue to be accepted as objective and impartial by the private and public sectors, it is understood that our fee for the undertaking of this project is in no way dependent upon the specific conclusions reached or the nature of the advice given in our report.

Other Matters

12 The client shall save and hold RERC free from claims that might arise in connection with this contract. Further, the client agrees to pay RERC for staff time, at RERC standard hourly billing rates, plus expenses at cost that might be required for expert testimony or any other court appearances, together with preparation time and legal costs that might arise because of RERC's involvement in this assignment and its familiarity with the properties, markets, and values associated with the assignment, whether subpoenaed by the client or any other group.

Questions for Analysis and Discussion

1. How many elements of a formal proposal can you identify in this letter? What features of proposals are emphasized here?

2. How does the first paragraph of the letter clearly establish the background information necessary to understand the rest of the proposal?

3. The writer of this proposal is careful to explain very clearly the terms of the project. Identify some instances in which this detailed explanation is particularly striking. Are there any points in the proposal that you think are ambiguous and might lead to a misunderstanding?

4. The attachment to the letter, explaining RERC's standard proposal terms and conditions, is appended to all RERC proposals. How does it supplement the information presented in the letter? What are the differences between the language of the attachment and that of the proposal itself?

Application

Write a letter proposing to perform a study or complete a project for your technical writing course. Indicate clearly the subject, focus, and scope of your proposed project. You should also describe your data sources, work schedule, and any other information necessary to make your proposal a persuasive document.

Michael D. Murphy,
John L. Motloch,
and Nan Booth Simpson

Request for Proposal;
Proposal

Three landscape architects, Michael D. Murphy, John L. Motloch, and Nan Booth Simpson (now associate garden design editor for *Southern Living*), prepared the following model RFP and proposal. The proposal is typical of the kind of document prepared by many consulting and design firms. The project described here involves the selection of an acceptable site for an elementary school and the design of the school's campus. Because the RFP describes a small project, the proposal itself is simple and is submitted in letter format.

1 September 1982

Request for Proposal

1 Recent growth in the community has placed increasing pressures on the Boulder Junction School District's physical facilities. Population growth is quickly outpacing the space in existing schools at all levels. Recent expansion efforts at the Middle and High Schools have dealt effectively with space needs at these levels for the present and the immediately foreseeable future. This, however, has not been true for the elementary schools. Expansions at these schools have not been able to bring them in line with desired space requirements.

2 The School District wishes to commence a school development program that will provide needed elementary school facilities in a way that assures that services will be provided when they are needed, where they are needed, both now and into the foreseeable future. Of equal importance is the School District's desire to assure that expenditures for its physical plant be spent in the best possible way consistent with high quality educational facilities.

3 Plans are to be developed that lead to the orderly and timely location and acquisition of suitable campus sites that best serve the educational needs of the community and are consistent with patterns of land use, circulation, population characteristics, and administrative control. This process must be capable of yielding adequate selection and development of elementary school sites throughout the growth phase of the community.

Project Phasing

4 **Part I**—A procedure must be developed that permits the School Board to identify and purchase needed school sites as the community grows and develops.

5 **Part II**—A site must be selected and developed to meet the immediate need for a new elementary-level school similar in size to those now in use, and in a location that is best suited to population character, access, and probable future use.

6 **Part III**—The selected campus site is to be designed in a way that maximizes educational and community opportunities, and minimizes costs, maintenance, and administrative difficulties. Possible future growth and expanded community uses should be considered in the design of the facility.

Proposal Deadline

7 Proposals must be received at the Superintendent's Office, Boulder Junction Independent School District, Boulder Junction, Michigan on or before 1:00 PM, 17 September 1982.

Proposal Format

8 One original and one copy typed single space on an 8½-by-11-inch sheet format.

Mr. Dennis R. Lowell, President
Board of Education
Boulder Junction Independent School District
100 Anderson Street
Boulder Junction, Michigan 49419

Dear Mr. Lowell:

1   Team Design Group and Associates (TDG) appreciates this opportunity
to present the following proposal to assist the Boulder Junction Independent
School District (BJISD) in planning elementary school facilities.

2   TDG will approach the project in three phases: 1) develop a procedure for
the selection of future elementary school sites, 2) employ that procedure to
assist the BJISD in selecting a site for a new elementary school facility, and
3) design the facility to maximize educational and community opportunities
while minimizing cost, maintenance, and administrative difficulties.

Phase I: Development of a Site Selection Procedure and Guidelines

3   TDG will formulate a site selection procedure in two parts that will enable
the School Board to identify and purchase school sites as the community
grows and to base site selection on established guidelines for evaluating potential sites according to appropriate location, spatial requirements, and program needs. During this phase of the project, TDG anticipates working
closely with school officials and agencies of the City of Boulder Junction. Research will be conducted to determine city growth patterns, population characteristics, present and future land use, availability of utilities, traffic patterns,
and other factors that affect the school district but do not fall within its control or jurisdiction. Additional research will focus on site-related factors, such
as grading and drainage, soils, and vegetation characteristics, in seeking to
predict and neutralize problems now existing on present school sites within
the city. TDG, with advice and data to be provided by BJISD, will also analyze and chart the service area and program needs of students, faculty and
staff, administration, and the community as a whole. From analysis of the
research data, a set of guidelines for site selection will be drawn. With these
guidelines TDG can assist BJISD in identifying potential problems and opportunities associated with sites to be considered for acquisition. By this procedure future problems may be adequately anticipated and trade-offs accurately predicted prior to the final site selection decision.

Phase II: Site Selection

4   TDG will apply the process outlined above to the selection of a suitable elementary school site in the section of the city identified as having the most

immediate need for a new campus. To assist the School Board in selecting a site that adequately fulfills the needs set forth in the guidelines, TDG will provide the following:

5   1. Target area designation—The portion of the city determined through comprehensive planning studies as presenting the greatest need for new elementary school facilities will be targeted for detailed site selection studies. Within the "Target Area" potential, available sites will be identified for evaluation of their suitability for acquisition and development.

6   2. Site analysis—Each potential site will be analyzed according to relations between the parcel and adjacent land uses, circulation and utilities. In addition, each site will be evaluated according to specific on-site factors such as spatial adequacy, topography, grading and drainage, soils, and vegetation.

7   3. Design evaluation—Each site will then be tested with preliminary schematic designs that employ established program criteria for evaluation and review by the School Board. These schemes will provide the basis for deciding the best site choice. With the data provided the School Board can make its final decision on the new campus location.

Phase III: Campus Design

8   *Part 1: Schematic Design*   Once the final site selection has been made, TDG will proceed with the design of the elementary school campus. To assure that the campus is designed as a single unit, TDG will direct a multidisciplinary design group, including architects, engineers, and landscape architects, in developing schematic design alternatives for the campus using programmatic data developed with the BJISD administration, faculty, and staff. Refinements to the design program will be made at this time to assure that the latest thinking on needs and requirements is incorporated into the design.

9   Design concept alternatives will be reviewed with the School Board. The schematic designs will be sufficiently refined to illustrate basic functional relationships within the campus such as building organization and locations, vehicular and pedestrian circulation patterns, service and maintenance, physical education and recreation areas, utilities, grading, and site drainage. Additionally, the general pattern of interior spaces, such as corridors, administration, classrooms, and cafeteria, will be shown for the buildings. Rough cost estimates will be provided for the schematic designs.

10   From these schematic design alternatives the School Board will be able to choose the concept it finds most satisfactory in meeting the needs of the BJISD.

11   *Part 2: Design Development*   Once the preferred design concept has been approved, TDG can begin the design development phase, in which concepts are refined and design dimensions, relationships and materials are developed

more fully. The design development phase will focus attention on specific elements—materials, costs, phasing, etc. in ways that permit the School Board to guide the development of the design within the framework of the overall concept.

12 ***Part 3: Construction Documents*** When the design has been refined and approved in sufficient detail, TDG will proceed with the preparation of construction documents. These include sufficient plans, details, specifications, and contract documents for bidding and construction of the project. The final package of construction documents will complete the design phase of the project. With these documents the BJISD will be prepared to advertise and accept bids for the work and enter into a construction contract with the selected building contractor.

13   Continuity and quality control throughout the construction phase of the project are important considerations. TDG could continue on the project in a construction supervision capacity at the pleasure of the School Board. TDG strongly advises this role to assure that the planning and design process are completed satisfactorily.

Time Schedule
14 Phase I will commence upon approval of this proposal and will be complete by 18 September 1983. Work on Phase II will begin immediately after completion of the site selection document and will end with selection of a site. Estimated date for the completion of Phase II is 2 October 1983. Phase III shall begin upon selection of a site and is expected to be completed by 4 December 1983.

Remuneration
15 The BJISD will compensate TDG as follows: All phases of the project will be billed an amount equal to two and one-half times the hourly rate of the planner's staff for all man hours worked plus cost of overtime premiums. The project director's time shall be billed at $30.00 per hour; staff at $15.00 per hour. It is recommended that the client initially budget approximately $5,000.00 for Phase I and $5,000.00 for Phase II.

16   In addition to the fees stated above, the following costs are reimbursable: Cost of maps, reports, photographs, site surveys and other documentation of existing base information necessary to the conduct of the work and not otherwise furnished by the owner; reproduction of drawings, specifications and other documents incurred for the project or requested by the owner except those for our own use; special presentation materials, printing, photographic enlargement/reduction and special art services as necessary to the preparation and production of reports; costs of surveys and test boring authorized in advance; fees, plus ten percent (10%) for special consultants authorized in advance.

Conditions of the Agreement

17   Original drawings will remain the property of TDG.

18   This proposal remains in effect for a period of ten (10) days. This agreement may be terminated by either party upon five (5) days' written notice. In the event of termination, TDG shall be paid compensation for services performed to termination date, including reimbursable expenses then due and all terminal expenses. In such an event, it is understood that we shall provide BJISD with a copy of all documents completed by us prior to our termination of work.

19   All payments due TDG shall be made monthly upon presentation of the statement of services rendered. Payments due the landscape architect under this agreement shall bear interest at the prime rate commencing thirty (30) days after the date of billing.

20   This document will serve as an agreement between us, and you may indicate your acceptance by signing in the space provided below and returning one (1) copy for our files.

21   We look forward to the possibility of working with the Boulder Junction Independent School District.

Respectfully submitted,

Michael D. Murphy/John L. Motloch
TDG and Associates.

ACCEPTED:

BOULDER JUNCTION INDEPENDENT SCHOOL DISTRICT

By: _____
       Dennis R. Lowell, President
       Board of Education

_____
       Date

Questions for Analysis and Discussion

1. How are the first two paragraphs of this letter effective in introducing the rest of the proposal?

2. Discuss and evaluate the extent to which the proposal responds adequately to the RFP. How do the organization and headings of the proposal letter correspond to the client's needs as defined in the RFP?

3. Both this and the SCS proposal (pp. 135–40) are proposals prepared by consulting firms for local governments. They differ, however, in format, length, and detail. How are these differences a function of the size and complexity of each project?

4. The writers of this proposal, like those of most successful proposals, state emphatically that certain things *will* be done. Why is this the most persuasive way of expressing future intentions?

APPLICATION

Much of the proposal is written in the passive voice. Would it be more effective in the active voice? Choose a section of the proposal and rewrite it in the active voice, using either *we* or *TDG* as the subject of the sentence.

Florida Game and
Fresh Water Fish Commission

Grant Application: Florida's Endangered Papilio Species

This unsolicited grant application is directed to the U.S. Fish and Wildlife Service and is submitted under the 1973 Endangered Species Act. Because it is unsolicited, part of its function is to convince the funder of the significance of the research. In its attempt to justify funding, the proposal clearly articulates the objectives of the study and the importance of the resulting information. Its straightforward language and clear pattern of organization help make this proposal a persuasive document.

■

Research Proposal

Study Title: Critical Habitat Determination and Recovery Plan Development for Florida's Endangered *Papilio* Species
Applicant: Florida Game and Fresh Water Fish Commission
            620 South Meridian Street
            Tallahassee, Florida 32301
Funding Agency: U. S. Fish and Wildlife Service Office of Endangered Species
Funding Source: Section 15, Endangered Species Act of 1973.

Background

1   Two rare papilionid butterflies, the Bahaman (*Papilio andraemon bonhotei* Sharpe) and Schaus (*P. aristodemus ponceanus* Schaus) Swallowtails, occur within Florida's Biscayne National Monument and portions of the Florida Keys (Brown 1973).* Both species were added to the federal endangered species list 28 April 1976 (FR 41:17740). Low abundance and a geographic distribution limited to extreme southern Florida have characterized both species since at least the 1930s (Bates 1934; Clarke 1940). Continued habitat destruction has fur-

---

*The references throughout and vita referred to on p. 162 are not included here because of space considerations.

ther reduced the geographic ranges of these species, largely restricting them to publicly-owned lands in the northern Keys (Covell & Rawson 1973; Brown 1973). Such range restriction heightens the possibility of extinction through chance natural occurrences (e.g., severe hurricanes or frosts). The geographic distribution of phytophagous insects is often limited by the distribution of suitable food plants. For *P. a. bonhotei* these include key lime *(Citrus aurantifolia)* and sour orange *(C. aurantium),* while *P. a. ponceanus* appears restricted to torchwood *(Amyris elemifira)* for larval development (Brown 1973). This specialization on a small number of plant species, however, does not explain the current range restriction of the butterflies. Their host plants are common throughout the Florida Keys and on the mainland, in areas where *P. a. bonhotei* and *P. a. ponceanus* no longer occur. While maintenance of these plant species plays an important part in the protection of existing populations, other factors clearly affect the butterflies' distribution within their potential geographic range.

Justification
2 The development of realistic Recovery Plans for these species will require greater knowledge of past and present distribution patterns and critical habitat requirements than is available in the current literature. Existing information is largely anecdotal, based on sporadic collecting trips by taxonomists and commercial dealers. This information, which is scattered through the literature dating to the 1890s, must be reviewed and analyzed if past distribution patterns are to be understood. Field studies are required to gain knowledge of critical habitat requirements and current geographic distributions.

Objectives
3 1. To determine past and present geographic distributions of the two species and their host plants within southern Florida.
   2. To define the species' critical habitat requirements and causes of the currently restricted geographic ranges, and to determine the factors affecting reproductive success in existing populations.
   3. To prepare a Recovery Plan for both species, including designations of Critical Habitat for each.

Procedures
4 Objective 1
   a. Accumulate, review, and analyze data from the existing literature to determine previous distribution patterns and their change over time.
   b. Survey areas known or strongly suspected to support these species in the past to determine presence or absence of current populations.

5 Objective 2
   a. Accumulate, review, and analyze existing data on past and present distribution of the species' host plants. Correlate plant distributions with insect distributions to determine the importance of host plant range on the insects' range. Determine the effects of other factors

(island surface area and plant species diversity, presence or absence of local freshwater sources and/or nectar supplies for adults, etc.) often found to limit phytophagous insect distributions.

b. Make on-site surveys to test hypotheses generated through the analysis of existing data.

c. Use nondestructive data-gathering techniques to determine the reproductive success of extant populations, identifying the causes and intensity of mortality in all life stages (egg-adult) and the mean fecundity of adults.

6   Objective 3

a. Interpret the resulting information and formulate a Recovery Plan. This plan will include designations of Critical Habitat and will outline steps to be taken in maintaining extant populations of these butterflies and encouraging their return to suitable areas historically included within their geographic ranges.

Schedule

	1980	1981
	J F M A M J J A S O N D	J F M A M J J A S O N D
Objective 1	X X X X X X X X X X	
Objective 2	X X X X X X X X X X	
Objective 3		X X

Costs

OPS Salaries	$15,108.00	($1259.00/mo. for Principal Investigator)
Expenses	2,400.00	(Computer time, travel expenses, etc.)
Equipment	600.00	(Collecting equipment, etc.)
Total	$18,108.00	

Personnel

Coordinator: Don A. Wood, Florida Game and Fresh Water Fish Commission, 904/488-3831

Principal Investigator: Gerold Morrison (vita attached)

Questions for Analysis and Discussion

1. What elements of this proposal indicate that it was written for a highly educated audience?

2. Look at the Procedures section of this proposal. In what ways does it demonstrate that the objectives of the study will be accomplished?

3. How do the format, style, and level of detail help make this proposal a more persuasive document? What questions remain in the reader's mind about the procedures the writer will follow?

4. One of the most important aims of this kind of proposal is to convince the reader that the research is valuable and worth funding. How does the writer achieve this goal? Point to specific aspects of the proposal that you think have helped the writer fulfill this objective.

Application

Select a problem of concern to students on your campus, such as parking, housing, food service, laboratory facilities, or computer availability. Using the headings of this proposal—Background, Justification, Objectives, Procedures—write a brief proposal requesting funding from the university or college for the collection and interpretation of data concerning this problem and for the formulation of a plan to correct it.

5

REPORTS

A report is a compilation of material written for a particular purpose and directed to a specific audience. For college and university students, the word *report* usually means term papers, complete with footnotes, bibliography, and long hours of library research. Term papers typically have a broadly defined subject as well as a narrowly restricted format. And although the avowed purpose of the term paper is to inform the reader, it is more often an exercise for the writer than a service to the reader. Moreover, the term paper is directed to the instructor, whose main purpose in reading such a report is to evaluate the student's mastery of a subject.

Term papers frequently offer a highly generalized treatment of their subjects. Such reports serve little purpose in the professional world, where the subject and scope of a report are strictly limited by the purpose for which it is written and the reader is critically concerned with grasping the information provided. Indeed, in the professional world, "reporting" can be more broadly understood as simply the collection and communication of information in response to a particular set of circumstances. This kind of report fulfills a specific need, answers a specific question, addresses a specific issue. Thus understood, reporting is perhaps the most important activity of the professional individual.

FORMAT OF THE REPORT

Just as the purpose and audience for which a report is written differ in every situation, so each report varies in the details of its format and execution. We can, however, distinguish informal from formal (or full-form) reports. The former are usually written as letters or memoranda and are generally designed as in-house communication or as communication between persons involved in a joint project. Informal reports are relatively brief and include a minimum of support material. Common examples include laboratory or field reports, accident or incident reports, reports of site visits or other business trips, and announcements of company policy, meetings, or changes. Sometimes printed forms are available to the report writer. In this case, the writer needs only to fill in the appropriate information. Accident reports, insurance claims, and income tax forms are examples with which most of us are probably unwillingly familiar.

Three examples of informal reports in memorandum format are included in this chapter. The first of these is an interoffice memorandum from the Lexington Public Power Supply System (pp. 174–77). Its purpose is to prepare recipients for participation in planning and budget meetings; its distribution is limited to those who will attend. A second example is the Monthly Progress Report from Arizona Public Service (pp. 180–85). Because this report is brief and will be distributed only among APS personnel, its informal format is fully appropriate. Had this report been prepared by one of the many outside companies involved in the described project, a more formal presentation would probably have been used. Finally, the Trip Report (pp. 188–90) illustrates one of the most common types of memo many professional employees have to write.

Longer and more complicated reports, prepared for an outside reader, use a formal or full-form format which is similar in many ways to that of the formal proposal. The formal or full-form report includes a number of elements (known as front matter and back matter) that support and clarify the body of information to be communicated. These elements can include

- *A Letter of Transmittal:* a letter from the writer of the report to the individual or organization for which it was prepared. This letter (sometimes called a "cover letter") accompanies the report, describes it briefly, and highlights any important points. Letters of transmittal also commonly accompany proposals and qualifications packages and serve the same introductory purpose (see p. 134, Chapter 4).
- *A Title Page:* a page that fully identifies the report by providing its complete title, the name or names of the individuals or organization responsible for its authorship, the date of submission, the person or organization to which the report is directed, and any other required identification data (such as document number, contract number, and so on).
- *Abstract or Executive Summary:* a brief, almost telegraphic summary of the report. The abstract either describes the scope *(descriptive* or *indicative abstract)* or summarizes the contents *(informative abstract* or *executive summary)* of the entire report. The indicative abstract, as its name suggests, indicates to the reader whether he or she should read the entire report.

In contrast, the informative abstract, to be useful, should be able to be read and understood independently of the report. It allows those who may not have the time or training to read or understand the full report to grasp its significance.

- *Foreword, Preface, Acknowledgments:* sections (which may be variously combined) that supply the background to the report. These sections describe the circumstances of the research—when, where, and for whom it was performed—and give credit to individuals, agencies, or institutions who have assisted in the performance of the research and the preparation of the report.
- *Table of Contents:* a page or pages giving an overview of the report by listing major (and sometimes secondary) headings of the report and their page numbers. The table of contents allows the reader to locate quickly those sections of the report in which he or she is most interested.
- *List of Illustrations or List of Figures and List of Tables:* a page or pages giving the full title and page numbers of all graphics.
- *List of Abbreviations, List of Symbols:* a page or pages giving the reader convenient access to the meaning of abbreviations and symbols used in the report, especially those that might be unfamiliar.
- *References Cited:* full documentation of all sources referred to in the text of the report.
- *Bibliography:* full documentation of all sources used in researching the report. An annotated bibliography also supplies a short description of each entry.
- *Appendices or Exhibits:* data that are relevant to the report but that are too long or complicated to be quoted in the body of the report itself.
- *Glossary:* a list defining any specialized terms used in the report

The article submitted for publication in a professional journal is a special kind of full-form report. The format specifications for publishable articles vary not only from field to field, but from journal to journal. Most technical articles will include an abstract and references section, and will use headings to highlight major divisions of the text. Acknowledgments may be included in the text of the article or in a footnote or endnote. Many professional periodicals also publish a capsule biography of the author or authors of the article, and some even include their photographs.

TYPES OF REPORT

The format and contents of any report will be particular to the situation or purpose for which it is written. There are, however, several general types of report that can be adapted for specific situations. Four common types are described below.

Progress Reports

A progress report provides information about the way in which a project is developing. This report is written primarily to give contractors, clients, supervisors, or anyone else who has a stake in the project an account of expenditures made, in terms of both time and money. Secondarily, it can help workers involved in the project to estimate the amount of work completed and work remaining to be done. Thus, the progress report offers both writer and reader an organized statement of the current status of a project.

A progress report may be submitted at regular intervals to provide a steady account of project development, as in the case of Arizona Public Service's (APS) Monthly Progress Report on its Particulate Removal Project (pp. 180–85). Or it may summarize the progress of a much longer period of time.

Progress reports can be organized in several ways. Some are structured chronologically, describing all work completed, in progress, or to be completed according to the time schedule of the project. Others list specific tasks and discuss how much work has been or needs to be done to achieve each task's objective. Another method is to organize progress reports by topic or area of responsibility, such as design, engineering, construction, and costs. Many progress reports, including the APS example, use a combination of these organizational methods.

Progress reports take many forms, depending on the frequency and formality of the submission. Some are prepared on printed forms. Short or in-house reports are usually written as letters or memoranda. Longer progress reports are bound and contain many of the elements of the full-form, formal report.

Other forms of periodically issued reports share many of the same goals and intentions with the progress report, but differ in content and organization. Monthly sales reports, periodic inventory reports, and analyses of a firm's financial status and management are common throughout business and industry. The quarterly and annual reports, produced by publicly traded corporations, also provide a regular assessment of a company's condition. Quarterly reports may be prepared as brochures; annual reports often look like glossy magazines. In short, an accounting of progress or analysis of present conditions is a useful tool common to most professional organizations.

Literature or State-of-the-Art Reviews

A literature (or state-of-the-art) review surveys the current status of research and development in a given area. It refers to and briefly describes works published on a particular subject during a certain period of time. The purpose of the literature review is to provide the interested and informed reader with a systematic review of all published material within the designated field. This field is usually quite narrowly defined, especially when the review is addressed to an expert audience, and is intended to be as comprehensive as possible. In addition to providing a complete catalogue of available literature, the review may also raise questions for new directions in research and indicate areas requiring further exploration.

A literature review often is included in the introductory section of research reports, technical articles, and theses and dissertations. Polich and Orvis' report on alcohol problems in the Air Force, excerpted in Chapter 1 (pp. 32–39), and Graber and Powers' study of "Dwarf Sumac as Winter Bird Food," included in this section (pp. 201–204), both incorporate a review of relevant literature. This review sets the context for the remainder of the study. It also guarantees that the researchers have a thorough grasp of their problem and are not repeating work that has been performed by others.

Literature reviews have no standard pattern of organization. Most begin with an introduction, which establishes the scope and significance of the review and may indicate its plan of organization. Key terms may also be defined as they apply to the study, especially when there is some controversy about their definition. Again, Polich and Orvis' review provides an example.

The body of the literature review is usually organized topically, according to the logic of the area surveyed. A typical technique is to offer a fact, observation, or opinion under study, followed by a catalogue and summary of relevant citations (references to published works). A conclusion is optional, for the literature review does not necessarily have to establish anything. If it is included, the conclusion may summarize the major points of the review and recommend areas or problems requiring further research.

Research Reports

A research report discusses research conducted by its author. The common vehicle of communication in the sciences and social sciences, as well as in technological research and development (R&D), this type of report may present the results of laboratory experimentation, field and site studies, surveys of public opinion and population trends, or any other variety of original research.

Certain reports—sometimes called *physical research reports*—describe direct research or experimentation conducted by its author. A report of this sort addresses a specific problem or question and, in response, advances a hypothesis—that is, a way of answering the question or solving the problem. The report then describes precisely and with scrupulous detail the method of experimentation the researchers have designed to test the hypothesis. Finally, the report presents and interprets the results obtained by the experimental procedure. The article "Dwarf Sumac as Winter Bird Food" (pp. 201–204), although brief, contains the major components of most physical research reports. These include:

- *An Introduction.* The introduction announces the problem, purpose, and scope of the study. It states the hypothesis to be tested and explains the scientific principles that led the researcher to erect that hypothesis. For this reason, past literature relevant to the problem will generally be reviewed in order to place the present work in context. The introduction may also describe the basic experimental design and mention any special equipment or evaluative methods employed by the researcher. In addition, the overall organizational plan will usually be included in the introduction to longer reports.
- *Materials and Methods.* This section gives a detailed account of experimental procedures. It contains a complete description of all materials used and all steps of the investigation. Also included are descriptions of laboratory procedures, data manipulations, and statistical analyses. The materials and methods of the investigation are ideally described with enough accuracy and precision to ensure that another investigator could, by reading this account, repeat the experiment exactly.
- *Results.* In this section of the report, the data obtained as a result of the research are presented in an organized manner. Frequently, data are displayed in tables, graphs, or other figures, to help the reader more easily assimilate a great deal of numerical or statistical information.
- *Discussion.* The discussion presents the researcher's interpretation of the results of the investigation in relation to the initial hypothesis or question. In short, it explains what the writer thinks his or her research shows. In this section of the report, the writer also evaluates the overall experiment or investigation in terms of prior research and discusses any flaws or errors in the design and execution of the project, and may raise questions for further research.
- *Conclusion.* The conclusion restates the essential points of the report and offers a succinct statement of what the study achieved. Recommendations for further research are sometimes included in this section.

Other research reports—sometimes called *investigative reports*—survey data, often collected by others, in order to describe the extent and scope of an existing situation. The researcher may simply present the data, or may analyze and interpret the data in order to draw conclusions. If the data are analyzed, the method of analysis, like the experimental procedure of the physical research report, must be carefully described.

Investigative reports contain most of the elements that make up physical research reports, including an Abstract, Introduction, Results, Discussion, and Conclusion or Recommendations. Often, however, rather than explaining experimental procedure, an investigative report will describe the source and methods of data collection, as well as any other factors influencing the data's reliability. An example of such an investigative report is "The New Media and the Demand for Studio Production Facilities" (pp. 191–200).

Feasibility Reports

A feasibility report determines whether a particular project can be performed with a reasonable degree of success. In other words, is the project practicable (can it be done) and practical (should it be done)? In making this determination, the feasibility report generally answers one or both of two questions:

1. Is a particular project desirable? That is, is it the best way—technically, economically, legally, environmentally, aesthetically—to tackle the problem?
2. Can the project be completed successfully in terms of time, money, available manpower, and any other relevant considerations?

The typical feasibility report contains a combination of five components: the introduction, the body, a summary, the conclusions, and the recommendations. The *introduction* of the feasibility report, as in most other types of report, outlines the purpose and scope of the study. It describes the problem to be addressed, together with the particular answers or solutions under consideration. The introduction may also offer a justification of the study, explaining why research needs to be done in this area. This is often accomplished by means of an historical survey of the problem—how it developed, why it needs to be solved.

The *body* of the feasibility report, divided into appropriate sections or chapters, presents the data collected in the process of examining the problem and its possible solutions. Information may include the results of site surveys, cost-benefit analyses, prototype design and testing, polling of public opinion, expert testimony—in short, anything that helps determine whether the project can and should be done.

Especially important elements of the feasibility report are the *summary, conclusions,* and *recommendations,* which may constitute separate sections or may be variously combined. The summary lists the most important facts accumulated during the study and the conclusions interpret these facts, again in light of the purpose of the study. Finally, the recommendations cite specific actions that should or should not be taken in order to respond best to the problem under study. These elements are so significant to the purpose of a feasibility study that they are often placed at the beginning, rather than the end, of the report.

Related to the feasibility report is the financial analysis, exemplified in this section by C. Marks Hinton's assessment of the Service Corporation International (pp. 207–15). Financial analyses are written for the potential investor who wants to decide whether to purchase stock or invest in an enterprise.

CONTENTS OF THE REPORT

Progress reports, literature reviews, physical research reports, and feasibility studies are four common types of report. The actual contents of a report, however, are as varied as the reasons for which it is written. Environmental Impact Reports (EIR) or Environmental Impact Statements (EIS), for instance, examine the effects on the human and natural environment of proposed development, and are closely related in form and function to feasibility studies. Planning and Development Reports map out and communicate to interested parties the future growth of a business, community, or institution. In short, anytime a need for information exists, a report—whether informal or formal—answers this need. Moreover, any individual project may require a variety of reports. And although most reports will conform to one of the patterns we have discussed here, the effective report writer will always be guided by the purpose, circumstances, and audience of each report he or she produces.

Benjamin Thomas
Lexington Public
Power Supply System

Memorandum: Budget Review

Memoranda are the principal vehicles for conveying any kind of information within an organization. They may be used to announce policy, report on field activities or attendance at professional conferences (as in the Trip Report, pp. 188–90), record progress (as in the Arizona Public Service Progress Report, pp. 180–85), or, as in this case, schedule meetings.

To be effective, memoranda must be both complete and concise. The reader should not have to guess the significance of the message nor wade through unnecessary words. This memorandum from Lexington Public Power Supply System effectively answers its recipients' needs. The first paragraph provides the context for the rest of the memorandum and previews the major subjects to be discussed. The logic of the message is carried out by the use of headings, lists, and parallel sentence structure. All elements of the communication—both formal and substantive—contribute to its cohesiveness.

Date: April 9, 1981   See Attachment
                                          BT/File/KL
   To: Distribution   ACB/File:ss

 From: Benjamin Thomas, Director of Administration

Subject: 1982 A&G/OWNERS' COST BUDGETS
         PLANNING AND BUDGET REVIEWS

The procedure for reviewing the 1982 Administrative and General/
Owners' Cost Budgets has recently been expanded to include working sessions aimed at improving the planning and budgeting process
and overall quality of the final plan. These working sessions are
scheduled for the period April 10 through April 17. Here is the
revised schedule for reviews:

        Detailed Planning and Budget Reviews....... April 10-17

        Joint Review of Program by Directors....... April 14

        Senior Management Reviews.................. April 29-30

Specific information about these reviews is given below.

I.   DETAILED PLANNING AND BUDGET REVIEWS

     Purpose: To (a) further develop the Supply System planning
     process, (b) improve integrity of the budgets, and (c) pro-
     vide summary comments, analysis, and recommendations to Senior
     Management for use in making budget decisions.

     Scope: The reviews will include all administrative and
     general, operating and maintenance, and owners' controlled
     construction costs budgeted during the budget period 1982
     through 1989. The programs relating to these budgets will
     become part of the 1982 Management Plan being developed by
     each Director.

Procedure: The reviews will be conducted by a Budget Review Committee, a working team of eight members representing Finance, Administration, Technical, Organization Performance, LNP-3, LNP-1/5, LNP-4/6 and Power Generation. Division Managers (or Directors) will present programs, staffing, and expenditure budgets for each Department-level organization to the Review Committee for review and discussion.

Objectives of the Budget Review Committee: To (a) review the planning assumptions of the programs, and the basis for their resource requirements, (b) identify duplications and omissions of planning among Supply System organizations, and (c) determine discretionary and non-discretionary programs and expenditures.

Preparation for Reviews: In preparing for the detailed reviews, managers should provide the following:

a. Completed "Programs and Manpower Staffing" forms (10 copies) for each department (sample form attached). These forms should include:

(1) program statement (title and purpose for each program);
(2) assigned staffing at June 30 of each year 1981 through 1989; and
(3) further breakdown of program staffing where directly related to projects (LNP-3, LNP-1/5, LNP-4/6, or general).

b. A BUD 670 budget report for each department (10 copies) listing budgeted expenditures by account and by year for 1982 through 1989.

c. Summary of budgeted owner-controlled construction costs by year (applies only to certain organizations).

Each manager should use whatever presentation approach and supporting information formats that are most effective for the organization. Please remember--these are working sessions, not formal presentations.

Locations and Schedule for Reviews: All reviews will be held in Room 3-108 of the Main Building (third floor, southeast corner). The attached schedule summarizes the established dates and times for the planning and budget reviews.

II. JOINT REVIEW OF PROGRAMS BY DIRECTORS--APRIL 14

Purpose: To review the current status and results of the
1982 A&G planning and budget process and to discuss any
identified or potential duplications of program responsibilities. This overview of the planning and budget process
will occur simultaneously with the bottoms-up detailed reviews being conducted by the Budget Review Committee.

Procedure: Directors will be briefed on the current status
of the A&G planning and budget process, the current budgeted
staffing and expenditure results, and any identified or potential duplications or uncertainties of program responsibility.

Preparation for Reviews: Directors should identify any
known or potential duplications or omissions in programs
and present these for discussion and clarification of
responsibility.

Location and Time of Review: Board Room, Main Building,
Tuesday, April 14, 9:00 a.m. to 11:00 a.m.

III. SENIOR MANAGEMENT REVIEWS--APRIL 29-30

Purpose: To present organizational budgets to the Managing
Director for review and approval.

Scope: All administrative and general, operating and
maintenance, and owner-controlled construction costs for
the budget period 1982 through 1989.

Procedure: Each Director will present the programs,
staffing, and budgeted expenditures for organizations
within his directorate to the Managing Director. Division-
level Managers may assist. All Supply System Directors
should be present for cross-communication of program and
budget information and for discussion as necessary. Specific procedures may vary, depending on the success of the
detailed budget reviews held earlier.

Preparation for Review: Specific requirements for these
reviews will be determined by April 22, 1981.

Location, Time and Schedule: The reviews will be held in
the Board Room of the Main Building, April 29 and 30,
8:00 a.m. to 4:00 p.m. each day. Specific agenda of presentations will be determined by April 22, 1981.

Attachments

DETAILED 1982 A&G/OWNERS' COST REVIEWS
SCHEDULE

	FRIDAY April 10	MONDAY April 13	TUESDAY April 14	WEDNESDAY April 15	THURSDAY April 16	FRIDAY April 17
8:00	-----	Engineering	Material Services & Contracts	LNP-4/6 Construction/ QA	Human Resources	Records Management
9:00	-----					
10:00	Internal Auditor	Technical	Procurement	LNP-4/6 Operations	Finance	Nuclear Safety
11:00	-----	Other Technical Services				Legal
12:00	LUNCH	LUNCH	LUNCH	LUNCH	LUNCH	LUNCH
1:00	Quality Assurance	LNP-3 Construction, QA	LNP-1/5 Const., Eng., QA, Pl., & Cont.	Power Generation Maintenance	Administration	Health, Safety & Security
2:00						
3:00	Corporate Information Systems	WNP-2 Operations	WNP-1/4 Operations	Power Generation Training		Public Affairs
4:00				Power Generation Services		Treasurer
5:00						

1982 A&G/OWNERS' COST BUDGET
PROGRAMS AND MANPOWER STAFFING

ORGANIZATION CODE: _____   ORGANIZATION TITLE: _____

PROGRAM/FUNCTION	(NUMBER OF EMPLOYEES AT JUNE 30)								
	1981	1982	1983	1984	1985	1986	1987	1988	1989

Questions for Analysis and Discussion

1. This memorandum contains instructions aimed at preparing personnel for effective participation in budget sessions. Locate and evaluate these instructions.

2. Instead of hiding them within the memorandum, Thomas states directly the purpose, scope, procedure, and other aspects of each budget meeting. What are the benefits to the readers of such explicit statements?

3. What features of layout and organization does the writer use to help the reader obtain the important information with a minimum of effort?

Application

Using a format similar to that of this memorandum, write an announcement of a forthcoming campus meeting, activity, or event. Use underlining, outlining, listing, or other forms of emphasis to highlight important information. ■

E. L. Lewis
Arizona Public Service Company

Monthly Progress Report

In 1980, Arizona Public Service (APS), a major utility corporation, began to expand its Four Corners power generating station by adding two units. To control emissions at this plant, APS implemented a number of features, including scrubbers and fan and baghouse devices. This memorandum records one month's progress on construction of emission control facilities at the two-unit expansion.

Although brief, the memorandum contains the important characteristics of most progress reports. First, the report begins with a summary that functions as an abstract, giving a concise digest of the overall report. Second, information is divided into major topics—Engineering, Construction, Cost Control, and so on. (In longer progress reports, these major sections are often distinguished with tabs, giving readers quick access to the subject of their concern or responsibility.) Additionally, headings are consistent from section to section, indicating work done, work in progress, and work remaining to be done. Moreover, topic division and headings—the format of the report—remain the same from one progress report to the next. Finally, information is presented as a list of tasks, not as great clumps of prose.

TO: C. D. Jarman
Sta. #   5579

FROM: E. L. Lewis
Sta. #   5188
Ext. #   7975

SUBJECT: Four Corners Units 4 and 5 Particulate Removal Project Monthly
         Progress Report—February, 1980

This is the February, 1980, Monthly Progress Report on project activities:

1. Pilot plant is still headed for 4/17/80 in-service date.
2. Finalizing plan and schedule to get 12/82 service date on full-scale
   project.
3. No change in cost estimates.
4. Accounting recorded charges of $452,000 in February.

A. Engineering:

1. Significant Milestone Accomplishments Achieved During the Month:
   a. Completed study on project fire protection needs.
   b. Met with Joy-Western, Wheelabrator-Frye, Research-Cottrell,
      American Air Filter, and Buell Envirotech to discuss filterhouse
      proposals.
   c. Provided recommendation and cost estimate to include restroom
      facilities in the new control rooms.
   d. Provided caisson survey notes to UE&C.
   e. Released bid package for construction yard.
   f. Reviewed the following UE&C specifications, studies, and
      documents:
      (1) Axial vs. Centrifugal Fan Study.
      (2) PDM Control Description.
      (3) Survey Services Specification.
      (4) Duct Fabrication Specification.
      (5) PRP and SO_2 Fan Study.
      (6) Demolition and Site Preparation Specification.
2. Areas of Work Involving Major Blocks of Manhours or Major
   Decisions:
   a. Flue gas acid dew point temperature evaluation.
   b. Evaluation of filterhouse proposals.
   c. Site survey for permanent brass caps.
   d. Review of demolition specification.
   e. Study of project fire protection needs.
   f. Release of bid package for construction yard.

3. Summary of Activities That Are Behind Schedule:
Relocation of construction facilities.

4. Summary of Key Activities Expected During the Coming Month:
   a. Complete comments on the following UE&C specifications and documents:
      (1) Site Preparation and Foundation Specification.
      (2) Concrete Specification.
      (3) Inspection and Testing Services.
      (4) Structural Steel Specification.
      (5) Structural Steel Erection Specification.
      (6) Drilled-in-Pier Specification.

B. Construction:

1. Significant Milestone Accomplishments Achieved During the Month:
   a. Completed demolition of horizontal scrubber test module.
   b. Received bids for relocation of construction facilities.
   c. Reviewed specification for demolition and site preparation.

2. Areas of Work Involving Major Blocks of Manhours or Major Decisions:
   a. Installation of baghouse test module.
   b. Demolition of horizontal scrubber test module.

3. Summary of Activities That Are Behind Schedule:
   a. Installation of baghouse test module.
   b. Relocation of construction facilities.
   c. Bid package for Unit 5 precipitator control room HVAC.

4. Summary of Key Activities Expected During the Coming Month:
   a. Continue erection of baghouse test module.
   b. Start work on relocation of construction facilities.
   c. Review bid package for Unit 5 precipitator control house HVAC.

C. Cost Control:

1. Significant Milestone Accomplishments Achieved During the Month:
Reviewed and approved engineering economic factors for use by UE&C.

2. Areas of Work Involving Major Blocks of Manhours or Major Decisions:
   a. Followed up on cost of construction and pre-operating power.
   b. Completed and routed draft of Contract Change Order Procedure.
   c. Started reauthorization of WA 19–2103 #5 Precipitator Upgrade.
   d. Obtained preliminary insurance data from UE&C.
   e. Analyzed January cost report.
   f. Updated Cash Flow Report.

Table 1 Four Corners Units 4 and 5 Particulate Removal Project*
Comparative Cash Flow Report (thousands of dollars)

| | APS | | UE&C | |
| | Current Month | Cumulative | Current Month | Cumulative |
1980	Est. Act. 0/ (U)	Est. Act. 0/ (U)	Est. Act. 0/ (U)	Est. Act. 0/ (U)
To 12/31/79	$ —	$ 323	$ —	$ 545
Jan.	194	517	148	693
Feb.	284	801	169	862
Mar.			186	1,049
Apr.				
May				
Jun.				
Jul.				
Aug.				
Sept.				
Oct.				
Nov.				
Dec.				

| | TOTAL PROJECT | |
| | Current Month | Cumulative |
1980	Est. Act. 0/ (U)	Est. Act. 0/ (U)
To 12/31/79	$ —	$ 868
Jan.	342	1,210
Feb.	452	1,663
Mar.		
Apr.		
May		
Jun.		
Jul.		
Aug.		
Sept.		
Oct.		
Nov.		
Dec.		

*Includes P.O.s 105–T1 and 105–K4 (230KV Power Supply)

g. Submitted new unit numbers to Budget Services for employee redistribution.
h. Corrected accounting for Lodge-Cottrell costs incurred by UE&C.
i. Prepared three-year cash forecast for Construction Budget Administration.
j. Reviewed UE&C procedure manual.

3. Summary of Key Activities Expected During the Coming Month: Update estimate to include cost of construction and pre-operating power.

Table 2   Four Corners Units 4 and 5 SO$_2$ Removal Project
Comparative Cash Flow Report (thousands of dollars)

1980	APS Current Month Est. Act. O/ (U)	APS Cumulative Est. Act. O/ (U)	UE&C Current Month Est. Act. O/ (U)	UE&C Cumulative Est. Act. O/ (U)
To 12/31/79	$ —	$ 37	$—	$—
Jan.	8	45	—	—
Feb.	68	—	—	—
Mar.				
Apr.				
May				
Jun.				
Jul.				
Aug.				
Sept.				
Oct.				
Nov.				
Dec.				

1980	TOTAL PROJECT Current Month Est. Act. O/ (U)	TOTAL PROJECT Cumulative Est. Act. O/ (U)
To 12/31/79	$ —	$ 37
Jan.	8	45
Feb.	68	114
Mar.		
Apr.		
May		
Jun.		
Jul.		
Aug.		
Sept.		
Oct.		
Nov.		
Dec.		

D. Insurance:

1. Significant Milestone Accomplishments Achieved During the Month:
   a. Strategy meeting with Project Manager, reviewing OCP approach, tentative implementation schedule and data needed from UE&C.
   b. Cost projections updated.

2. Summary of Key Activities Expected During the Coming Month:
   Upon receipt of PDM and other UE&C data, present underwriting data to Brokers.

E. Procurement:

1. Significant Milestone Accomplishments Achieved During the Month:
   a. Reviewed specifications 20,100,313; 20,100,314; 20,100,315; 20,100,316; 20,100,317.
   b. Received and distributed copies of bids received on Inquiry No. 20,100,310 (500KVA transformer).
   c. Sent Inquiry #20,100,308 (Booster Fans) to prospective bidders.
   d. Made commercial evaluation of bids received on Inquiry No. 20,100,300 (Baghouse).
   e. Issued the following purchase orders for the Filterhouse Pilot Plant:

P.O. #	Vendor	Description
20,100,499	Elder Leasing	Mobile office
20,100,800	Morgan Drive Away	Move mobile office
20,100,801	Patent Scaffolding	Field supervision
20,100,802	Morgan Drive Away	Move mobile offices
20,100,803	Leeds & Northrup	Charts
20,100,804	Riley Co.	Relays & flasher
20,100,805	G.E. Supply	Motor starter heaters
20,100,806	Eastside Electric	Fuses

F. Filterhouse Pilot Program:

1. Significant Milestone Accomplishments Achieved During the Month:
   a. Coordinated construction and engineering work by APS.
   b. Evaluated contractor bids for FPP testing.
   c. APS low temperature design work.
   d. Selection of bag material and bag manufacturer.
   e. Coordinated LCD and S/R engineering.
   f. Operating Manual—80% complete.

2. Areas of Work Involving Major Blocks of Manhours or Major Decisions:
   a. All of the above.
   b. Responding to field engineering problems.
   c. Reviewed UE&C technical and commercial evaluations.

3. Summary of Key Activities Expected During the Coming Month:
   a. Complete operator's manual.
   b. Sign contract with FPP test contractor.
   c. Determine requirement for low temperature FPP approach and method to achieve that.
   d. Track erection of FPP unit.
   e. Track I&C and electrical FPP checklist.

G. Scheduling—
Filterhouse Pilot Plant:

1. Significant Milestone Accomplishments Achieved During the Month:
   a. "Tie-ins" for both Units 4 and 5 completed.
   b. Module area support towers installed.
   c. *Pre-Ass'y* ductwork from module area to Unit 5 completed.
   d. Module area scaffolding engineering completed.

2. Summary of Key Activities That Are Behind Schedule:
   a. Installation of scaffolding from Unit 5 to Unit 4.
   b. Valve control wiring diagram (APS Engineering—Jan. 25 was late finish)
   c. Module area support details.
   d. Module area scaffold installation.

3. Summary of Key Activities Expected During the Coming Month:
   a. Begin duct insulation.
   b. Ductwork installation—Module to Unit 5, Unit 5 to Unit 4.
   c. Instrumentation installation.

H. Project Agreement:

1. Significant Milestone Accomplishments Achieved During the Month:
   a. Rescheduled two-day meeting from February to March 18 and 19. These two days should bring us to the final form of the Project Agreement with possibly only the final economic items remaining.
   b. A report of the meeting will be delivered orally on March 20th to E. L. Lewis by R. W. Gimbernat and E. B. Dellaquila.

I. Other:

Attached is a copy of UE&C's Monthly Progress Report for February, 1980.

> E. L. Lewis, Project Manager
> Emission Abatement Projects

ELL/iec
Attachment

cc: T. Woods   E. Chartier
    L. Mundth   E. Jochim
    A. Simko   W. Ekstrom
    B. Clark   Field Office
    F. Heacock   File: 000.1
    D/F: 3/21/80   204.3

Questions for Analysis and Discussion

1. Compare the language and organization of this memorandum report with those of a more formal report, such as "Dwarf Sumac as Winter Bird Food" (pp. 201–204) or "The New Media and the Demand for Studio Production Facilities" (pp. 191–200). Why has Lewis chosen to issue this memorandum in outline rather than paragraph form? Is his use of sentence fragments legitimate?

2. This report lists very specifically accomplishments, activities behind schedule, and expected work. What are the advantages, to both writer and readers, of this attention to detail? Why would it be unacceptable simply to say "Work is proceeding as planned" or "We're a little behind schedule"?

3. What features of this progress report indicate that it is written for readers who understand the nature and scope of the work described? Choose one major section and explain what specific information would need to be added, were the memo directed to an audience unfamiliar with the project.

4. Compare this memorandum with the budget review memo (pp. 174–77). What common features of style, organization, and word choice can you discover? On the basis of these memoranda, can you suggest some general guidelines for memo writers?

Application

Using this memorandum as a model, write a memo reporting your progress toward completion of a major project you are working on for your technical writing or other course. Then rewrite that same memo using paragraph rather than outline format. What are the relative advantages and disadvantages of each?

Trip Report

The Vector Electronics Company is a small corporation specializing in the manufacture and sale of circuit boards and related electronic components. The following trip report was written by employee F. L. Hill to provide Vector's president with an account of the company's participation in a regional trade show.

The organization and language of this memo are typical of most in-house reports of trips, attendance at meetings or conventions, or other events of company concern. After providing a brief introductory summary, the writer describes in detail his meetings with present and potential customers, using headings to indicate his organization by topic and numbered lists to set off individual items of information. The straightforward, conversational tone of this report reflects the daily working relationship between company personnel and their shared knowledge of the business and concerns for its prosperity.

To: D. Scoville
cc: R. Scoville

Call Report—F.L. HILL

Date: October 22, 1982

To: Simcona Electronic Show
Maplewood Party House
Rochester, NY

3:00 P.M. to 9:00 P.M.

1   I arrived at the Electronic Show after making calls at Simcona in Rochester, and set up the display around 2:30. Approximately 2,200 people who had been invited by Simcona attended the show. Simcona had provided colored badges that identified participants as engineers, buyers, purchasing agents, or technicians. Our booth was well attended, allowing me to field several questions and problems as well as develop potential new business. The following points developed during the show:

A. Problems

2 1. Freda Peck, the buyer at Simcona, reported that the University of Rochester had notified her of a problem on our 8803-6-1 motherboards, had returned their board, and did not want to reorder. She didn't know if this was because they had already purchased another board. Ms. Peck said Frank Meyers, one of their sales people, reported they were losing quotes, but she had no more specific information.

3 2. Bob Baxter (telephone: 716-555-2467) complained about the general workmanship in the cages and modules sent to him. He also complained that delivery was too long and that he had repeatedly asked for drawings without receiving them. We are to phone him during the week of October 24–27 to get more specific information.

B. Quote Requests

4 1. John F. Zima, Senior Engineer
Engineering Division
Bldg. # 23,
Kodak Park Division
Eastman Kodak Company
Rochester, NY 14650   (telephone: 716-555-2412)

Zima requested a quote for 10 to 12 cages. His requirement is sketched on the attached RFQ. He also requested copies of the layout sheet on our 3677–2, 4112–4, and 3682–6.

5   Zima also had a 4½" × 6½" board with 22/44 contacts and an unusual layout pattern for prototyping. It used a grid design with spaces for DIPS, about 1 DIP high by 2 DIPS wide, running across the width and length of the board. This board, which did not bear the vendor's name, was custom-designed. Eastman Kodak wants another board like it, and Zima will send us a sample to quote up to 1000 pieces as a stores item. This board was a little different from anything we have, and might be one we'd want to consider constructing. None of the standard boards I showed Zima in the catalogue seemed to him as useful as this one.

C. Product Requests

6   1. I had several requests for multibus motherboards. Eastman Kodak uses the multibus system and now employs 20-slot motherboards made by Electronic Solutions.

7   2. I had another request for a cage into which the Electronic Solutions motherboard could be dropped to simulate an Intel cage (model IC5180). This cage has four fans on the bottom and a reset button, and is widely used at Eastman Kodak. We have seen this cage before and should investigate it further. Additional business may become available by providing Vector Pak cages for motherboards and daughterboards too large to be plugged into racks.

D. Potential Customers

8   1. Steven C. Kunnmann, Product Manager
       Board Products
       Gordos Corporation
       250 Glenwood Avenue
       Bloomfield, NJ 07033   (telephone: 201-555-6800)

This company, which also exhibited at the show, makes products for the STD bus and does not have a motherboard or module-holding system. Kunnmann believes that his electrical maintenance customers would like housing around the STD bus boards. He considered the EFP case and the cage I had on display as possible answers. I will follow up this contact with STD bus information at their Bloomfield, NJ, address.

9   2. A.J. DiPasquale, Manager
       Purchasing and Logistics
       Redcom Laboratories, Inc.   (telex: 555-498)
       750 Fairport Park
       Fairport, NY 14450   (telephone: 716-555-0390)

Redcom has been purchasing cages from Bud that do not meet their requirements because of what they feel to be poor quality control. Specifically, there was so much slop in the card guides that the boards could pop out. DiPasquale had other complaints as well, and I said we'd like a chance at their business. He promised to send bids to us. They are now buying 100 cages a year that use a smaller board and 1000 cages a year that use a bigger board. This sounds like a serious request, which we'll follow up.

Questions for Analysis and Discussion

1. Identify the differences in the relation between writer and reader (or readers) that are apparent in the style and content of this memo and the budget review memorandum (pp. 174–77). Why is a less formal, more conversational style appropriate in this report? Would that style be acceptable in the budget review announcement?

2. What features of this selection show that the writer could assume that he and his reader shared a great deal of common knowledge? Do these same features make it difficult for you to understand this selection? What additional information would you want the writer to supply?

3. During the six hours the writer was at the Simcona Electronics Show, he met a number of people with different problems and needs. How has he organized the material describing his various encounters? Why is this organization more effective than simply listing, in chronological order, everyone he met?

Application

Imagine that you are the executive for whom this report was written. Write a response to this memo, explaining your reaction to the information and requesting the actions you wish taken. Be specific.

J. N. Dertouzos and
K. E. Thorpe

The New Media and the Demand for Studio Production Facilities

The informative or evaluative research report relies upon published material and other sources to provide information needed for an understanding of a particular subject. Such reports usually address a specific question, as in the following example. This report examines the changes brought about in the television entertainment industry as a result of expanded video technology and attempts to predict the effects these changes will have on the demand for production facilities. Authors Dertouzos and Thorpe depend upon a variety of sources for their data and carefully document this information. However, their conclusions, which are based upon their evaluation of the data, are their own.

1   The entertainment industry is in the midst of dramatic changes in its economic environment. New technologies and a relaxed regulatory atmosphere are likely to stimulate a period of growth unequaled since television's inception in the early 1950s. As a result, an increased demand for cable programming and the associated impacts on existing media will no doubt strain the capacity of firms to provide sound stages for the creation of new products. This report will examine these changes and assess the probable consequences for the program supply industry as well as the derived demand for production facilities.

The Rental Market For Sound Stages

2   With the growth of television, the movie industry began to decline along with the fortunes of major studios. The demand for stage facilities decreased significantly. This situation was aggravated by the changing technology of feature production. That is, the development of compact cameras and mobile equipment, as well as the increased use of videotape, created a severe excess capacity in stage production facilities. More and more shooting was done on location. As a result, a competitive rental market for studio space developed.

3   For the most part, stage space has been readily available from major studios. Independent producers have had easy access to stage facilities and entry into the industry has been uninhibited. Indeed, economic evidence suggests that program supply has been extremely responsive to slight shifts in demand [1]. Surely the ease of entry has been due, in large part, to the excess capacity in production facilities. This has served to encourage the increasing role of small, independent suppliers. Such firms, with small and volatile shares of the production business, depend on efficient access to the capital provided by the rental market.

4   The importance of independent producers and their increasing participation can be seen in data describing prime time programming shares for 1970 and 1978. To illustrate, Table One describes the market share distribution of action, adventure, and dramatic series. During the eight-year period, the number of such programs increased from 27 to 60. During the same time, the number of suppliers increased to 33, almost double the total eight years ago. Clearly, the market shares of the largest firms have been falling. Perhaps most significantly, the share of dramatic programming for the six major film studios has fallen from 61.5 percent to 50.3 percent[2]. Table Two displays identical information for the supply of prime time comedy series. From 1970 to 1978, there was a 35 percent increase in the number of comedy programs broadcast. Again, the share distribution indicates that the largest firms are becoming less dominant. The share of the major film studios has fallen rather sharply. In 1970, the major film studios supplied 41 percent of the prime time hours devoted to comedy. By 1978, their share had fallen to only 26 percent.

Table 1   Supply of Network Programs:
Prime Time Action-Adventure/Dramatic Series

Distribution Among Firms	Market Shares (percent of total hours)	
	1970	1980
Leading Firm	27.9	28.8
Leading Four Firms	55.0	51.5
Leading Eight Firms	75.1	69.4
Leading Twenty Firms	100.0	94.8
Major Studios	61.5	50.3
Total Number of Series	27	60
Total Number of Suppliers	18	33

Source: Federal Communications Commission, Network Inquiry Special Staff, 1980.

[1] See, for example, Crandall (1970) and Owen, Beebe, and Manning (1974). Also, for a description of the program supply industry in earlier years, see Arthur D. Little, Inc. (1966).

[2] For these data, the major studios are considered to be Universal, Twentieth Century Fox, Paramount, MGM, Warner, and Columbia.

Table 2   Supply of Network Programs: Prime Time Comedy Series

Distribution Among Firms	Market Shares (percent of total hours)	
	1970	*1980*
Leading Firm	12.4	9.7
Leading Four Firms	46.5	37.4
Leading Eight Firms	66.6	66.6
Leading Twenty Firms	100.0	99.1
Major Studios	41.0	26.4
Total Number of Series	32	43
Total Number of Suppliers	20	23

Source: Federal Communications Commission, Network Inquiry Special Staff, 1980.

5   Telephone interviews with several production studio staff members confirmed the historical existence of significant excess capacity[3]. Although the recent collective bargaining problems have resulted in a short-term decline in studio utilization, the unanimous opinion was that stage space is becoming more and more difficult for independent producers to rent. Before the strike, sound stage use had been running at between 85 and 100 percent of capacity. Expectations are that future demand will easily exceed current levels. Space is so tight in some studios that there have been internal conflicts over the optimal allocation of facilities.

6   The market changes can be linked directly to the increased production schedules of the major studios. Before the recent work stoppage, a dramatically rising number of movies were being planned for release. For example, MGM, Warner, MCA, and Twentieth Century were increasing combined movie production by an aggregate of 60 percent over a one-year period[4]. The impetus for these increased production plans was the demand for programming by cable networks as well as the rising use of movies for prime time television. Increased demands for movies have been reflected in astounding rises in the prices earned for both cable and television rights. Fees have more than tripled over the last year[5]. The increased demand for studio space will certainly escalate in the near future. The major cable companies such as HBO and Showtime have already started to scramble for programming. As the major studios take up more and more of their sound stage space to meet this demand, the independent producers, with lower priority in the queue for rental facilities, are bound to suffer.

---

[3] Discussions were held with officials of seven major studios, including all those that do significant business with independent producers. The studios contacted included Universal City Studios, TBS Burbank Studios, Zoetrope Studios, Producers Studio, Laird International Studios, MGM Studios, and Walt Disney Productions.

[4] *The Wall Street Journal,* June 17, 1980.

[5] *Dun's Review,* July 1980.

Technology, Regulation, and the Growth of Cable

7 The programming boom has been fed, in large part, by the growth of the cable television industry. Table Three documents the dramatic increase in cable penetration. As of 1980, 15 million homes had cable television, about 20 percent of the total. This represents a steady rise since 1955. Recent estimates place the figure at just over 27 percent, representing a remarkable rise of about 35 percent in less than a year[6]. Since 1975, the penetration has more than doubled and is expected to surpass the critical threshold of 30 percent this year, a level which, media experts maintain, will establish the industry as a viable advertising medium as well as a source of entertainment.

8 The recent cable boom is not surprising given recent relaxation of Federal Communications Commission regulations. As late as 1972, several restrictions served to inhibit the growth of cable, especially in major markets[7]. In particular, in 1976 the FCC ended the "leapfrogging" provision which required that systems in large markets import signals only from nearby cities. In 1977, the pay cable rules that restricted the use of feature films were eliminated. In addition, cable networks were, for the first time, allowed to bid on programming usually shown on broadcast channels[8]. Next, public access provisions were overturned, thereby freeing more channel space. Last year, the FCC finally eliminated remaining restrictions on distant signal carriage and exclusivity rules that barred the use of syndicated material for one full year after such programming had been sold anywhere in the United States. The upshot of these regulatory changes is that the infant cable industry can now achieve its vast potential.

Table 3 The Growth of Cable Television Industry

	Number of Systems	Number of Subscribers
1955	400	150,000
1960	640	650,000
1965	1325	1,275,000
1970	2490	4,500,000
1975	3506	9,800,000
1980	4225	15,500,000
1981*	4300	21,000,000

*Estimated

Source: Television Factbook, 1980.

[6] Television Digest, September 7, 1981.

[7] There are several excellent summaries of FCC regulations and the role they played in stifling cable growth. The best of these include Besen and Crandall (1981), MacAvoy (1977), and Rosse and Dertouzos (1978).

[8] The motive for these regulations was the protection of traditional broadcasting networks and outlets. These policies were misguided attempts to serve the public interest.

9   Technological changes have also played a pivotal role in the development of cable television. First, in 1977, the FCC authorized the use of 4.5 meter satellite earth station receivers. This development vastly altered the economics of distant signal importation and has prompted rapid satellite development. Such transmission technology has changed the role of cable from one of signal enhancement to new program supply. There now exist nine domestic satellites each with about 24 transponders. The current shortage of space will probably be ameliorated with the tripling of capacity by 1984[9]. The expected boom in programming demand is documented by the fact that over 95 percent of the future transmission capacity approved by the FCC is already leased or purchased.

10   Even though the entertainment industry has already been affected by the growth in cable, it is clear that the impacts on program supply and producers are only just beginning. Transmission costs via satellite are less than five percent of the costs associated with telephone and microwave transmission. Also, the cost of earth stations has fallen by an average annual rate of over 12 percent over the last six years[10]. Finally, the emergence of over 20 cable networks and program brokers has further lowered the costs of reaching the growing numbers of cable subscribers. It is no small wonder that firms in several diverse industries have established cable programming divisions. It is a glamour industry with a spectacular future.

The Future of the New Media and Impacts on Program Demand

11   The rise in cable penetration and programming activity has been rapid indeed. However, the future is still quite uncertain. Industry projections are understandably speculative and must be interpreted cautiously. It is clear, however, that past studies have severely understated the ultimate reach and impact of this medium[11]. It is certain that the current level of penetration, just over 27 percent, significantly understates the future penetration of cable. Penetration has exceeded 70 percent in some small markets. Most larger cities, as they become wired, will contribute to the aggregate penetration level. Table Four [p. 196] indicates the rather slow growth experienced in the cable penetration of the top 20 markets. As suggested earlier, this is probably due to

---

[9] *Business Week,* September 14, 1981.

[10] FCC Network Inquiry Staff (June 1980).

[11] Early studies predicted that cable would ultimately penetrate anywhere between 30 and 50 percent of television households. See, for example, Park (1972). However, Crandall and Fray (1974) showed that, for individual markets, these models imparted a significant downward bias and, in fact, failed to predict even current levels of cable penetration. These failures certainly limit their credibility in viewing the future. Their failures are not surprising, however. To begin with, the econometric models were applied to data that were generated in an economic environment marked by limited programming options, restrictive regulations, and extremely high costs of transmission. Thus, past estimates of cable's potential can no longer be taken seriously. However, they are useful in that they suggest that the recent estimates, though more optimistic, remain based upon market assumptions that are also likely to become obsolete as the industry matures and technological change slows.

the more severe FCC restrictions imposed on systems in these markets. The average penetration in these markets is a mere 13 percent, significantly below the industry average. Cable penetration in Chicago, Detroit, Washington, Dallas, St. Louis, Houston, Miami, and Baltimore is, on average, below five percent. These markets sum to 20 percent of all television households. Clearly, the wiring of these markets will have a significant impact on the overall importance of cable television.

Table 4 Major Market CATV Penetration, Fall 1979

Market	Television Households (000)	CATV Penetration (%)
New York	6,398	15
Los Angeles	4,051	15
Chicago	2,850	3
Philadelphia	2,399	20
Detroit	1,600	2
Boston	1,807	12
San Francisco	1,884	30
Cleveland	1,349	16
Washington	1,398	9
Dallas	1,175	7
Pittsburgh	1,137	35
Houston	1,104	6
St. Louis	980	3
Miami	944	6
Atlanta	936	11
Minneapolis	994	5
Tampa	889	16
Baltimore	807	2
Seattle	915	25
Indianapolis	745	17
Top 20 Markets	34,362	13

Source: Television Factbook, *1980.*

12   Table Five [p. 197] presents data on market penetration projections for a variety of broadcast media and electronic accessories likely to have impacts on the demand for programming. Cable television is expected to penetrate about one out of two households by the end of the current decade. As suggested earlier, this estimate ignores the potential impact of developing technology, falling transmission costs, and limitless role of ancillary services being created for cable. Much higher levels of penetration are more than a possibility. In addition, major independent television stations like WGN in Chicago and WOR in New York will penetrate additional markets. Household penetration will, according to the projections, achieve 80 percent by 1990. With a larger potential audience for independent stations, they will be able to compete more effectively with the network affiliates, potentially creating a demand for higher quality programming. The rise in pay cable, from under 8 percent to 35 percent by 1990, will have a dramatic impact on the type of programming demanded. Revenues per viewer will far exceed the few cents

Table 5 Projections of Broadcast Penetration, 1979–1990

	Household Penetration	
	1979	*1990*
Network	100	100
Independent Stations	71	80
Public Broadcasting System	90	92
Cable	20	50
Pay Cable	8	35
Subscription TV	1	8
Direct Broadcast Satellite	—	5
Video Cassette Recorder	2	17
Video Disk	—	28
Telex View Data	—	33

Source: RCA Electronic Business Development Committee, Television Digest, *September 7, 1981.*

per television household that advertisers currently pay for audiences at prime time[12]. Also, the FCC has lifted restrictions previously imposed on subscription television (STV). Penetration is expected to increase in the future, especially in areas not served by cable. The prediction for direct broadcast satellites must be interpreted with care since there exists no experience with their operation. There remain questions concerning the costs of transmission. However, as the technology develops and costs become manageable, new and exciting outlets for programming may develop. Finally, the markets for video cassettes and disks are expected to expand. Prices have already started to fall and market penetration is expected to be significant in the not too distant future.

Competitive Media and the Threat to Broadcasting

13   It is clear that consumers will have considerably more program choices in the future. Not only will more households have access to cable, but channel capacities are also increasing. Table Six [p. 198] illustrates the growth in capacity of existing CATV systems between 1977 and 1979. The increase in the largest category of channel capacity was by far the greatest. Newest technologies permit almost unlimited channels. There will be no dearth of outlets for creative programmers.

14   What impact will these changes have on traditional broadcast television? As suggested earlier, the FCC policy to restrict cable was largely geared towards protecting free commercial television. Policy makers feared that audience fragmentation would make advertiser supported broadcasting less viable. However, several studies suggest that the networks will continue to be profitable despite some degree of audience diversion to cable[13].

[12] One need only witness the prices charged for receiving recent prize fights.
[13] See Park (1979), for example.

Table 6   Channel Capacity of Existing CATV Systems: 1977, 1979

Channels	1977	1979	Percent Change
13+	966	1219	+ 26
6–12	2759	2793	+ 1
1–5	176	151	− 14
Total	3901	4163	+ 7

Source: Television Digest, *1978 and 1980.*

15   Although the networks may withstand the cable onslaught, the changes will necessitate some competitive responses. For one, network packagers will have to extend beyond the 26 weeks of the current seasons for first-run programming. In the past, broadcasters have taken advantage of audience inertia. That is, aggregate television audiences are, for the most part, unresponsive[14]. People watch television regardless of whether or not reruns or first time programming is on. As long as all three networks operate under the same calendar, they are all better off having short seasons, thereby reducing programming costs[15]. However, with increased competition from cable, networks will have to supply better quality and first run programming all year around to avoid significant losses to cable. Thus, production companies will have to supply more and better programming to networks in the face of cable competition.

Conclusions

16   Several developments in the entertainment industry suggest that the future holds vast promise for the suppliers of programming. Recent technological advances in program transmission and reception have created a multiplicity of outlets where there once existed only a few. The relaxed regulatory atmosphere has permitted this technology to advance rapidly after years of stagnation. Cable development is exceeding even the most optimistic of predictions. In addition, the competitive response of the networks will likely create further demands for creative program development. At the same time, the boom in programming will affect rental markets for studios. Sound stage facilities, once inflicted with severe excess capacity, are already becoming fully utilized. Independent producers who are supplying an increasing percentage of prime time programming will be the first to be squeezed out unless studios are developed to meet the growing demand.

[14] Noll, Peck, and McGowan (1973).

[15] In essence, industry standards and conventions are implicit (or explicit via the National Association of Broadcasters' Codes) methods of collusion to maximize the profits of a few oligopolists.

Bibliography

Baer, Walter, and Carl Pilnick. *Pay Television at the Crossroads,* The Rand Corporation, P-5159, April 1974.

Barnett, Harold J. "Perspective on Cable TV Regulation," in W. G. Shepherd and T. G. Gies (eds.), *Regulation in Further Perspective: The Little Engine that Might,* Ballinger, Cambridge, 1974.

Besen, Stanley, and Robert Crandall. "The Deregulation of Cable Television," *Journal of Law and Economics,* Winter 1981, pp. 79–124.

Branscomb, A. W. "The Cable Fable: Will It Come True?," *Journal of Communication,* Vol. 25, No. 1, Winter 1975.

Brotman, Stuart N. "The New Era," *The Wilson Quarterly,* Winter 1981, pp. 76–85.

"Cable Revolution," *Consumers Research Magazine* 63, October 1980, pp. 11–13.

Charles River Associates, Inc. *Analysis of the Demand for Cable Television,* Report No. 78-2, Cambridge, Massachusetts, April 1973.

Committee for Economic Development. "Broadcasting and Cable Television: Policies for Diversity and Change," April 1975.

Crandall, R. W. "The Postwar Performance of the Motion Picture Industry," *Anti-Trust Bulletin,* 1970.

Crandall, Robert. "Regulation of Television Broadcasting: How Costly Is the Public Interest?," *Regulation,* January/February 1978.

Crandall, Robert W. and Lionel L. Fray. "A Reexamination of the Prophecy of Doom for Cable Television," *The Bell Journal of Economics and Management Science,* Vol. 2, No. 1, Spring 1971.

"Expanding Cable and Pay TV Networks Inject New Life into Movie Industry," *Wall Street Journal,* June 17, 1980, p. 27.

Federal Communications Commission Network Inquiry Special Staff, Final Report. *New Television Networks: Entry, Jurisdiction, Ownership and Regulation,* October 1980.

――――. *An Analysis of Television Program Production, Acquisition, and Distribution,* June 1980.

"The Gold Rush of 1980," *Broadcasting,* March 31, 1980, pp. 52–56.

"Hollywood Battles for New Markets," *Dun's Review,* July 1980.

"How Cable-TV Success Hinges on Satellites," *Business Week,* September 14, 1981, pp. 89–90.

"Hughes Spacecraft to Serve Cable Television Market," *Aviation Weekly* 114, June 15, 1981, pp. 54–55.

"In Scramble to Bring Cable TV to Your Area," *U.S. News* 89, October 6, 1980, pp. 47–48.

Arthur D. Little, Inc. *Television Program Production, Procurement, and Syndication,* Vols. I and II, FCC Docket 12782, 1966.

MacAvoy, Paul W., ed. *Deregulation of Cable Television,* American Enterprise Institute, Washington, D.C., 1977.

Manning, Willard, and Bruce Owen. "Television Rivalry and Network Power," *Public Policy,* Vol. 24, No. 1, Winter 1976.

Noll, R. G., M. J. Peck, and J. J. McGowan. *Economic Aspects of Television Regulation,* Brookings Institution, Washington, D.C., 1973.

Owen, Bruce, J. Beebe, and Willard Manning. *Television Economics,* Heath, Lexington, 1974.

Owen, Bruce, and Ronald Braeutigam. "Regulation of a New Technology: Cable Television," *The Regulation Game,* Ballinger, Cambridge, 1978.

Panko, R. R., G. C. Edwards, K. Penchos, and S. P. Russell. *Analysis of Consumer Demand for Pay Television,* Stanford Research Institute, Stanford, California, 1975.

Park, R. E. *Audience Diversion Due to Cable Television: A Statistical Analysis of New Data,* The Rand Corporation, R-2403-FCC, April 1979.

"The Race to Feed Cable TV's Maw," *Fortune,* May 4, 1981.

Rosse, J. N., and J. N. Dertouzos. "Economic Issues in Mass Communications Industries," *Proceedings of the Symposium on Media Concentration,* Vol. I, Bureau of Competition, Federal Trade Commission, December 1978.

Subcommittee on Communications of the House Committee on Interstate and Foreign Commerce. *Cable Television: Promise Versus Regulatory Performance,* 94th Congress, 2d Session (Subcomm. Print), 1976.

Questions for Analysis and Discussion

1. Reread the objectives of this report as set out in the introductory paragraph. How well does the report succeed in meeting these objectives? Now write an informative abstract that effectively summarizes the report.

2. Look at the list of sources upon which this report is based. Discuss the variety and relative reliability of these sources. What effect does this bibliography have on your willingness to accept the writers' conclusions?

3. Examine this report to distinguish between the writers' opinions and the facts upon which those opinions are based. Have the authors provided enough data to prove that their conclusions are correct? Should they have provided more information in any areas?

4. How effective is the use of tables in this report? Could the information contained in these tables have been presented in a different (verbal or graphic) form? Would it then be easier or more difficult to grasp?

Application

Select a current issue, problem, or topic of discussion in your major field of study. Develop a working bibliography of recent sources on your topic. Gather sources from as many different kinds of publications as possible: journal articles, books, government documents, company literature, popular press, and so on. Be sure not to neglect microtext and computer-assisted information retrieval services.

Jean W. Graber
and Pamela M. Powers

Dwarf Sumac as Winter Bird Food

Research reports describe the scope, procedures, results, and significance of scientific investigation. Such reports are commonly written by researchers in most sciences and social sciences and in some fields of engineering. This article from *The American Midland Naturalist* is a well-written example of a physical research report. It avoids generalizations, employs concrete language and specific terminology, pays close attention to detail, and organizes the information in a logical manner. Its organization is typical of most reports of original research and includes an Abstract, Introduction, Materials and Methods, Results and Discussion, Acknowledgments, and Literature Cited sections. Note that the conclusion is integrated into the section titled Results and Discussion.

Journals publishing research reports of this kind usually specify the format they wish authors to follow. These format requirements are designed to keep the report concise and economical and to exclude anything not strictly essential to the reader's understanding of the scope and significance of the research. They also guarantee that documentation will be standard, enabling interested readers to judge the validity of the reported research and locate published accounts of related studies.

1   **Abstract:** Birds consistently chose fruits of dwarf sumac *(Rhus copallina)* rather than of smooth sumac *(R. glabra)* when both were available as winter food. Dwarf sumac had larger fruits and higher caloric value than smooth sumac.

Introduction

2   Treatises on nutritional values of wildlife food plants (*e.g.,* King and McClure, 1944; Martin *et al.,* 1951) have indicated that "sumac," though eaten by a number of bird species, was used only as an emergency food when other foods became scarce. Errington (1937) found that captive pheasants failed to

maintain their body weight when fed only sumac. However, in these references the sumac being considered was either smooth sumac *(Rhus glabra)* or staghorn sumac *(R. typhina).*

3   In past winters in southern Illinois we noticed that birds often fed on sumac. In the autumn and winter of 1979–1980 we observed that birds were heavily using dwarf sumac *(Rhus copallina)* but were not using smooth sumac even when both species were growing side by side. We saw several species of birds eating dwarf sumac, including downy woodpecker *(Picoides pubescens),* blue jay *(Gyanocitta cristata),* Carolina chickadee *(Parus carolinensis),* American robin *(Turdus migratorius),* bay-breasted warbler *(Dendroica castanea),* rufous-sided towhee *(Pipilo erythrophthalmus),* slate-colored junco *(Junco hyemalis)* and white-throated sparrow *(Zonotrichia albicollis).* Birds made use of dwarf sumac from September to March. By March most of the dwarf sumac fruit had been picked, but the fruit of smooth sumac showed little sign of use. After the dwarf sumac drupes had been largely picked, some fruit began to disappear from the fruiting heads of the smooth sumac.

In E-central Illinois where dwarf sumac is absent and smooth sumac is common we have seldom seen birds eating sumac. Curious about the decided preference of birds for dwarf sumac, we examined fruits of both dwarf and smooth sumac more closely.

Materials and Methods

4   Entire fruiting heads of dwarf and smooth sumac were collected in old-field habitat near Golconda, Pope Co., Illinois, on 7 February and 23 March 1980. Fruit samples came from 10 different plants of each species in the same 5-sq-km area. Samples of drupes were removed as carefully as possible, weighed, dried to constant weight in a hot-air oven and refluxed in petroleum ether for 48 hr. Other samples were dried, compressed into approximate 1-g pellets and burned in a bomb calorimeter. Weights of clean, undamaged individual drupes were determined by counting the number of fruits in approximately 6-g samples. Six-g samples were also used for fat extractions.

Results and Discussion

5   It was immediately evident that dwarf sumac fruit contained more oil than smooth sumac fruit because our hands became "greasy" when removing the dwarf sumac drupes from the fruiting heads, but not when we handled the smooth sumac drupes. In fact, oil could be seen on the surface of the dwarf sumac drupes. Upon extraction we found that fruit of the dwarf sumac had a crude fat (ether-soluble extract) of 14.7%, whereas that of smooth sumac had 11.0%. Reports in the literature indicated even higher fat content for dwarf sumac than we found. The percentage crude fat based on dry weight in dwarf sumac was 17.7 (King and McClure, 1944) and over 25 (Short and Epps, 1976), while that of smooth sumac was 10.0 (King and McClure, 1944), 11.2 (Wainio and Forbes, 1941), and 14.2 (Bonner, 1974). In none of these studies were the samples of the two sumacs taken from the same locality.

6   In the process of making pellets with a hydraulic press for burning in the bomb calorimeter, much oil was lost from the dwarf sumac samples and some from the smooth sumac. Caloric values of dwarf and smooth sumac samples averaged 5013 cal/g and 4763 cal/g, respectively, using this procedure. We then placed whole drupes in the calorimeter expecting that when ignition took place some of the material would be blown out of the cup, but this did not happen. The dwarf and smooth sumac drupes averaged 5301 and 5152 cal/g, respectively, with this procedure. The difference in calories between the two species in our small sample (n = 7) is barely significant (t = 2.32, p = <.10), but when calories per gram are converted to calories per fruit—84.7 for dwarf sumac and 61.2 for smooth sumac—the increased caloric efficiency for birds using dwarf sumac becomes apparent. There are about 28% more calories per fruit (*i.e.,* per bite) in dwarf sumac than in smooth sumac, based on average fresh weights of 0.0173 g per fruit and 7.74% moisture content for dwarf sumac and 0.0128 g per fruit with 7.59% moisture for smooth sumac.

7   Brizicky (1963) found that the histological character of the fruit pulp (mesocarp) varied from species to species in the genus *Rhus,* and was quite different in smooth sumac from some other sumacs. He did not mention dwarf sumac, but he did describe the mesocarp of poison ivy *(Rhus radicans),* which is another much-used winter bird food. The mesocarp of poison ivy was filled with "wax," which consisted mainly of glycerides of fatty acids—mostly palmitin but also some olein and a small amount of free, dibasic high molecular fatty acids. We suspect that the mesocarp of dwarf sumac may likewise consist largely of fatty acids.

8   Fruits left on dwarf sumac heads in March were 41% lighter than those collected in February (averaging 0.0102 g per fruit). Birds presumably had left the less desirable drupes. Fruit might have been lost from the fruiting heads because of wind action or other mechanical action. To test this we tapped a number of fruit-bearing stems of both smooth and dwarf sumac. Fruit was very easily knocked off smooth sumac heads, but much less easily from those of dwarf sumac, yet the smooth sumac had retained most of its fruit. From our numerous observations of birds feeding on dwarf sumac, and from the way the fruiting heads looked after fruit was removed, it seemed fairly certain that birds had removed most of the fruit that was taken.

9   Dwarf sumac offers considerably more caloric value per drupe as a winter food than smooth sumac. Smooth and staghorn sumac have been the species considered when authors have suggested that sumac is not a choice winter food for birds. Dwarf sumac may be more important as an avian food than has been thought, even being used by a migrant species, the bay-breasted warbler, which could have had little or no experience with dwarf sumac on either its breeding or wintering grounds.

10   *Acknowledgments.*—We wish to thank Richard R. Graber and Ronald E. Duzan for their help in the laboratory.

Literature Cited

Bonner, F. T. 1974. Chemical components of some southern fruits and seeds. *U.S. For. Serv. Res. Note SO-183.* 3 p.

Brizicky, G. K. 1963. Taxonomic and nomenclatural notes on the genus *Rhus* (Anacardiaceae). *J. Arnold Arbor. Harv. Univ.,* 44:60–80.

Errington, P. L. 1937. Emergency values of some winter pheasant foods. *Trans. Wis. Acad. Sci. Arts Lett.,* 30:57–68.

King, T. R. and H. E. McClure. 1944. Chemical composition of some American wild feedstuffs. *J. Agric. Res.,* 69:33–46.

Martin, A. C., H. S. Zim and A. L. Nelson. 1951. American wildlife and plants: a guide to wildlife food habits. McGraw-Hill, New York. 500 p.

Short, H. L. and E. A. Epps, Jr. 1976. Nutrient quality and digestibility of seeds and fruits from southern forests. *J. Wildl. Manage.,* 40:283–89.

Wainio, W. W. and E. B. Forbes. 1941. The chemical composition of forest fruits and nuts from Pennsylvania. *J. Agric. Res.,* 62:627–35.

Jean W. Graber and Pamela M. Powers, Illinois Natural History Survey, 172 Natural Resources Building, Urbana 61801. *Submitted 27 May 1980; accepted 11 July 1980.*

Questions for Analysis and Discussion

1. How have the writers integrated past studies into their report? What functions do these references serve?

2. Scientific and technical writing is often jargon-ridden and wordy. This report, however, has a direct and economical style, partly because the authors use the first person *we* and the active voice. Rewrite the following exaggerated example of poor scientific writing (or another such example you have found in your own reading) to make it more simple, direct, and effective:

   A number of samples (5) were then placed within the experimental solution, upon which event it was expected that their colors would change. This occurrence having failed, it was hypothesized that the strength of the solution was not strong enough to effect the expected transformation.

   Discuss the major differences in the clarity and economy of the two versions.

3. Although this article is an objective account of a scientific experiment, it does not completely hide the writers' personalities. The authors describe the curiosity that motivated them, their responses to certain occurrences, and their reactions to their research. Find specific instances where you think the character of the writers is revealed. What is the effect of these personal notes?

4. The findings of much scientific and technological investigation are reported in professional journals. These journals have strict requirements regarding format, style, manuscript preparation, and documentation. Select a journal in your field and examine its instructions to authors. Compare the presentation of an article in your journal with that of this article from *The American Midland Naturalist.* Find out if documentation practices are standardized in your field.

Application

Abstracts of articles from professional journals are collected in publications such as *Biological Abstracts, Engineering Abstracts,* and *Psychological Abstracts.* Find out what abstracting services are available in your field and select an abstract from one of them. Then locate and read the article to which it corresponds. Does the abstract accurately reflect the contents of the article?

C. Marks Hinton, Jr.
Underwood, Neuhaus & Co.

Financial Analysis of Service Corporation International

The following financial analysis of the Service Corporation International (SCI) was prepared by C. Marks Hinton, Jr., of Underwood, Neuhaus & Co., a Texas brokerage firm. The purpose of this report is to inform the potential investor of the status of the company and to advise that investor of the relative risks and benefits of purchasing its stock. In other words, the report addresses the investor's question, "Should I buy or sell?" Brokerage firms provide such analyses as a continuing service to their clients. Companies that trade stock publicly, such as SCI, themselves prepare quarterly and annual reports for their shareholders. Also, any corporation whose annual income exceeds $10 million must submit a report to the Securities and Exchange Commission of the U.S. government. The picture of the company reflected in each of these reports—the investment analysis, the quarterly and annual reports, and the 10K report—can differ substantially, in accordance with the differing audience and purpose of each.

This financial analysis resembles a feasibility report in form and intention. Both kinds of report strive to assist the reader to make a decision. Thus, the reader is principally interested in the results of the writer's investigation. For this reason, such reports begin with a summary of the conclusions that have been reached on the basis of the research, and give a prominent place to the resulting recommendations.

SERVICE CORPORATION INTERNATIONAL

1 Recent Price: $17   1980 Price Range: $19–$7 ⅛
   Dividend: $0.44   Yield: 2.6%
   Earnings Per Share: (1)   Price/Earnings Ratio:
           FY1979: $1.63
           FY1980: $2.04   FY1980: 8.3x
       Est. FY1981: $2.70–$2.80   Est. FY1981: 6.1x–6.3x
   Shares Outstanding: 2,870,846   Traded: NYSE
   Estimated Float: 1,700,000 Shs.   Symbol: "SRV"

   (1) Fiscal year ends April 30.

Summary and Conclusions

2 ***Recommendation*** Purchase of Service Corporation International's common
stock could prove rewarding for risk-oriented investors seeking capital appreciation over the longer term. However, recent strength in these shares limits
near-term attractiveness.   ■

3 ***Rationale***
   1. Management's five-year strategic plan calls for doubling the size of the
      Company. This growth is expected to come from acquisitions,
      diversification, and internal operating gains.
   2. Increases in Service Corporation's pre-arrangement funeral revenues,
      expansion of its floriculture operations, and an expected rise in the
      nation's death rate should allow income to grow more rapidly in the
      coming five years than was recorded in the 1976–1980 period.
   3. The large amount of urban real estate that the Company is carrying on
      its balance sheet appears to represent an undervalued asset. (The
      American General Insurance transaction discussed on page 7 of this
      report indicates just one profitable use of these properties).

The Industry

4 ***An Overview*** The funeral/cemetery industry is highly fragmented. With
the exception of two sizable funeral home chains and several lesser-sized
units, most of the nation's 22,000 facilities are family-owned and operated.
This latter group accounts for over 90% of the industry's annual receipts. The
industry has experienced steady, albeit modest, growth in recent years. Total
sales have risen from $1.5 billion in 1967 to an estimated $4.0 billion in 1980,
for a compound annual rate of 6.3%.

5   The nation's death rate has been declining from its recent peak in 1967 of 9.7 deaths per thousand persons to 8.8 per thousand in 1978, the lowest level in U.S. history. This action has been the result of an overall increase in the health consciousness of the populus combined with certain strides made in the medical community with relation to various forms of cancer and heart disease. Nonetheless, demographic patterns indicate a reversal of this trend should begin occurring in the early-to-mid 1980's, as today's large senior citizen population moves past the age of 75.

6   In the past, the funeral industry has come under the scrutiny of the Federal Trade Commission (FTC) for alleged violations of the FTC Act. Investigations have centered upon allegations that certain funeral homes have overcharged for their services, misrepresented state laws concerning the handling of cadavers, falsely represented products, and made illegal payoffs to medical examiners to aid funeral homes in receiving business. Although little proof has been uncovered to support this myriad of charges, the investigation has resulted in some negative publicity for the industry, especially for the large chains that were the most visible targets for governmental attacks. In 1978, the FTC proposed that Congress approve certain regulations that would require funeral homes to: 1) itemize prices of a service's various components; 2) give quotations of price via telephone; 3) require permission of the closest relative prior to embalming the deceased; and 4) display inexpensive caskets as well as costly ones. While these proposals were not accepted by the Legislature, the Commission continues to review the industry's practices and may again seek Congressional support for some type of regulation in future years. In our opinion, however, quality funeral directors would welcome any restrictions that would improve the industry's image and put a halt to unscrupulous operators.

7   *Operations*   Most funeral homes offer a complete package of services, including embalming, transportation, supervision of services, and required paperwork. Cremation or burial, cemetery plots, vaults, monuments, flowers, and clergy costs are generally extra.

8   Funeral costs (excluding the items listed above as extra) vary tremendously. Burial cooperatives advertise prices as low as $295, while the price of an elaborate service can easily top $10,000. The average for the most recent year in which data is available (1977) was $1,412. This average compares to $1,027 three years earlier.

9   From an economic standpoint, the funeral industry is characterized by the operating leverage associated with most service industries. The necessity of handling peak-load business requires a considerable investment in fixed assets (land, buildings, equipment, vehicles) and personnel (embalmers, funeral directors, 24-hour answering service). Thus, while breakeven levels are high, once this case load barrier is reached, small changes in volume generally produce relatively greater increases in net income. In addition, the industry has two other positive economic virtues: 1) Funeral homes have a high cash flow, with accounts receivable usually being paid off within 30 days from life in-

surance proceeds or estate settlements; and 2) Dunn and Bradstreet rates the funeral industry as the group with the lowest failure rate of any industry classification in the U.S.

10 *Future Outlook*   Over the long term, the outlook for the funeral industry is favorable. With an aging population, the death rate is expected to begin rising once again in the 1980's. The U.S. Department of Commerce projects this total will rise to 9.6 per thousand by 1984 vs. 8.8 in 1979. The increasing acceptance of pre-need arrangements for funerals should also be a positive factor in the industry's future growth.

11 On the negative side, investors should be aware that the number of cremations has been increasing in recent years. Although nationwide cremations account for only 9% of all funerals, in the Pacific Coast States this total has risen to 30%. Economics of cremation result in lower revenues to funeral operators. In addition, the industry will probably face more competition from "non-profit" burial societies in coming years.

The Company

12 *Principal Business*   Service Corporation International (SCI) is the largest funeral home/cemetery operator in the nation with approximately 3.0% of the total market. The Company owns and operates 190 funeral homes, 29 cemeteries, 17 crematories, and 30 flower shops. SCI's operations are conducted in 21 states, the District of Columbia, and five Canadian provinces.

13 The Company was incorporated in Texas in 1962. Much of SCI's growth has come from the firm's aggressive acquisition program. During its most active phase, 1968–1973, the Company purchased over 100 funeral homes and some of its cemeteries and crematories. SCI's venture into the flower business began in 1978, as management sought a way to diversify into a semirelated industry. (This move is of interest from two points of view. First, more than 40% of domestic flower sales are made in conjunction with funeral services—providing SCI with a "natural" vertical expansion of its operations. Second, the flower business, like the funeral industry, is highly-fragmented, usually family-owned, and generates a high cash flow.)

14 *Company Structure*   Service Corporation's operations are divided into three groups. We will discuss the specifics of each.

15 **1. Funeral Homes** (1980 revenues: $96.5 million; operating profit: $24.1 million—83.9% and 89.3%, respectively, of total consolidated results.)

16 At the end of fiscal 1980, SCI owned and operated 184 funeral homes. During the first quarter of the current year, the Company acquired four additional units and completed construction of another home.

17 The funeral, crematory, and floral operations are divided into five geographic areas, each of which is administered by a regional vice president. Specifically, these divisions are as follows (1980 revenues): 1) Eastern: $29.1 mil-

lion generated from 46 funeral homes and crematories and two flower shops; 2) Southeastern: $18.1 million generated from 41 funeral homes and crematories and four flower shops; 3) Southern: $18.2 million from 26 funeral homes and crematories and six flower shops; 4) Central: $19.1 million by 33 funeral homes and crematories and eight flower shops; and 5) Western: $16.0 million from 38 funeral homes and crematories and five flower shops. (Geographic totals include $4.0 million from floral operations.)

18   During fiscal 1980, SCI acquired six funeral homes. While these new operations contributed to the $11.8 million rise in revenues, over half of this increase came from existing homes. At established locations, gains in case volume as well as price increases were responsible for the improvement in sales.

19 **2. Cemeteries** (1980 revenues: $14.5 million; operating profit: $2.2 million—12.6% and 8.1%, respectively, of total consolidated results.)

20   Although small in comparison to the Company's funeral home revenues and income, cemetery operations nonetheless remain an important part of SCI's overall corporate structure. Like the funeral operations, cemeteries are divided into regions also. At yearend, the breakdown was as follows: 1) Eastern—no cemeteries; 2) Southeastern—$1.9 million from six units; 3) Southern—$1.4 million from five locations; 4) Central—$4.5 million from seven cemeteries; and 5) Western—$6.7 million from eleven units. This group has shown rapid growth in recent years due to improved management procedures and a more aggressive cemetery acquisition program.

21   For the fiscal year ended April 30, 1980, cemetery revenues rose by $5.6 million. Of that gain, $5.2 million can be attributed to the six cemeteries acquired in FY1980 and the thirteen purchased the previous year.

22 **3. Flower Shops** (1980 revenues: $4.0 million; operating profits: $0.7 million—3.5% and 2.6%, respectively, of total consolidated results.)

23   Management chose to enter the floriculture business in 1977, believing, as stated earlier, that it was a logical expansion of SCI's funeral operations. Since that time, the Company has proceeded cautiously in this area in order to be certain of the potential that flowers seem to offer. Two approaches have been taken. The first was to establish flower shops within existing funeral homes in order to serve them and nearby units. To date, there are 13 in-house shops. The second was to purchase existing full-service flower shops that served not only SCI's operations but other retail accounts as well. There are currently 17 of these locations.

24   During FY1980, the Company acquired seven new flower shops and has added five more in the current fiscal year.

25 *Future Outlook*   While funeral home operations will continue to provide the largest contribution to SCI's revenues and income during the foreseeable

future, we believe the cemetery and floricultural activities will show the most rapid growth. In addition, the Company's pre-need marketing activities should become a more important source of profits. Currently, the pre-arrangement sales are conducted in only 5 (Texas, Florida, Missouri, Colorado, and Washington) of the 21 states in which SCI operates. As more state legislatures recognize the benefits of pre-need contracts to the consumer, we expect that legislation may be enacted making this type of transaction available to more purchasers. In the five states now served, the dollar value of funeral pre-need programs increased by 33% to almost $20.0 million in fiscal 1980.

26   Management's long-term strategy calls for doubling the size of SCI over the next five years. While we view this as an ambitious goal (revenues would have to compound at a 14.9% rate compared to 9.5% in the 1976–1980 period), we approve of the assumptions that underlie the growth plan. Specifically, the program itself contains six features: 1) Expansion of existing funeral operations through additional acquisitions; 2) Purchasing cemeteries in those cities in which SCI has existing funeral facilities or plans entry; 3) Construction of funeral homes in cemeteries owned by SCI or the purchase of funeral home/cemetery combinations in major market areas; 4) Aggressively promoting the Company's pre-need programs in existing and new state markets; 5) Expansion of SCI's floriculture operations via acquisition and internal construction; and 6) Diversification into a non-funeral related business. Management has discussed this possibility in the past, and we believe they will continue to review new acquisition candidates in the future.

Financial Data

27 *Operating History: 1976–1980*   (See Table I.) During the five-year period 1976–1980, revenues at Service Corporation increased from $80.1 million to $115.0 million, for a compound annual gain of 9.5%. During the same period, net income and earnings per share rose from $5.0 million ($1.30 per share on a fully-diluted basis) to $7.4 million ($2.02), compound increases of 10.5% and 11.6%, respectively. On a 10-year basis (1971–1980) the gains are more impressive, with revenues compounding at 14.6%, net income at 17.9%, and per-share earnings at 17.7%.

28   During the periods listed above, SCI reported one year (1977) of lower earnings. For the fiscal year ending April 30, 1977, net income fell to $4.0 million from $5.0 million a year earlier. Several factors contributed to this decline, including lower sales at existing funeral homes, a drop in cemetery revenues as the result of the sale of some units, a non-recurring gain in 1976 from the sale of real estate, and a change in the amortization period of a covenant against competition.

29   For the year ending April 30, 1980, the Company's return on average equity rose to 12.4%, from 10.9% twelve months earlier.

30 *Recent Results*   For the first quarter of fiscal 1981 ending July 31, Service Corporation reported the following results:

Table 1   Service Corporation International First Quarter Earnings, Fiscal 1981

	First Quarter Ending July 31 ($000)		
	1980	1979	% Change
Revenues	$31,445	$26,288	+19.6%
Net income	2,010	1,632	+23.2
E.P.S. (Fully diluted)	0.54	0.45	+20.0

31   These earnings were slightly in excess of our expectations. Comparisons were being made against a very strong three-month period in fiscal 1980. While SCI's funeral operations reported good gains, a very strong showing by the cemetery division was a key factor in the rise in income.

32 *Revenues and Earnings Projections*   For the fiscal year ending April 30, 1981, our projections of revenue and income are as follows:

Table 2   Service Corporation International Projected Revenues and Earnings, Fiscal 1981

April 30 ($000)	1981 Estimated	1980 Actual	% Change
Revenues	$ 129,500	$ 115,009	+12.6%
Net Income	8,700	7,404	+17.5
E.P.S.	2.73	2.02	+35.1
Estimated Dollar Variance	2.70–2.80	None	N.A.

33   These projections are based upon the following assumptions:

1. The anticipated increase in revenues comes equally from acquisitions and internal growth;
2. Continued focus on pricing, cost controls, and expense reductions combine to improve margins;
3. No major diversification is made outside of SCI's traditional funeral-oriented businesses;
4. The tax rate approximates 46%; and,
5. The approval of the American General Life Insurance stock redemption program (discussed later in this report) reduces the average number of shares outstanding during the period to 3.2 million compared to 3.6 million in FY1980.

34 **The American General Transaction** Shareholders of Service Corporation approved a plan to redeem 709,461 shares of its common stock from American General Insurance Company in return for property SCI owns at the corner of Sage and Westheimer in Houston's Galleria area. The acreage is currently occupied by the Geo. H. Lewis & Sons Funeral Home. A major attraction of the transaction to Service Corporation was the non-taxable nature of the exchange.

35 As structured, SCI traded the property, which was carried on its books at $1.6 million, for $11.0 million of its common stock and incurred no tax liability on the property's appreciation. (The $11.0 million for the five-acre site valued the land at $50.50 per square foot, a value our Real Estate Department believes is realistic. The ¾-acre site at the Northeast corner of Sage and Westheimer recently sold for approximately $75.00 per square foot.) SCI did give American General a concession stating that if the latter sold the land within three years for less than $11.0 million, Service Corporation would contribute the difference up to a maximum of $3.0 million. SCI will continue to lease the property from American General for two years at an annual rate of $150,000. At the end of the leasehold period, the Company will move the Geo. H. Lewis operations to another location.

36 The transaction also eliminated the convertible feature of the $1.0 million in SCI debentures owned by American General. The modification terms did not include an upward revision of the interest rate.

37 This event brings us to another aspect of Service Corporation that is more difficult to quantify, but should be considered by investors. As of April 30, 1980, the Company was carrying $48.3 million in real estate on its books, at cost. As this property is spread throughout 21 states, Washington, D.C., and Canada, we have been unable to ascertain an accurate current market value of these assets. While we feel it unrealistic to expect the value of all this real estate to have appreciated by the magnitude of the Geo. H. Lewis property (about 600% over its cost) since the Galleria area is certainly a special situation, we do believe these assets are worth more than they are being carried in the balance sheet. If this assumption is correct, SCI represents an interesting asset play on the underlying value of its real estate.

38 **Summary Financial Position** (See Table II.) Based upon Service Corporation's latest audited balance sheet (April 30, 1980), we rate the Company's financial condition as satisfactory. The current ratio was 2.0 to 1, and working capital totaled $15.4 million. It should be noted that $13.5 million of SCI's equity is represented by goodwill. Taking this into consideration, tangible net asset value per share is $13.80 compared to a stated yearend book value of $17.59. (Both of these figures will increase on the current year's audited numbers due to the reduction in the number of outstanding shares as a result of the American General exchange.)

39 SCI's capital structure consists of $68.5 million in long-term debt (46.7%), $15.2 million in deferred items (10.4%), and $62.8 million of equity (42.9%). The debt-to-equity ratio stands at 1.3 to 1. While this ratio might appear

quite high, it should be noted that the majority of the Company's debt is self-liquidating, in that it is directly related to the acquisition of the funeral homes and cemeteries, and repayment is more than covered by those entities' cash flows. Over 80% of SCI's debt is at fixed rates that average less than 8%; therefore, the Company has had little exposure to the recent violent fluctuations in interest rates.

40 **Dividend Policy**   Service Corporation commenced the payment of cash dividends in 1974. To date, the Company has paid 28 consecutive quarterly dividends to shareholders. The payout has increased each year since 1975.

Although the Board of Directors has announced no formal dividend policy, it has stated an intention to increase dividends to the extent possible within the limits of sound business practices. In the past, the payout ratio has approximated 17%.

41 **Stock and Options Ownership**   Robert L. Waltrip, Chairman of the Board and Chief Executive Officer, is the largest shareholder of Service Corporation, with 504,189 shares (17.6% of the outstanding common). Other officers and directors hold 455,831 shares (15.9%). Institutions report ownership totaling 301,700 shares (10.5%).

42   As of April 30, 1980, the latest fiscal yearend, there were options outstanding covering the purchase of 120,700 shares at prices ranging from $3⅞ to $5¼.

Technical Information

43 **Stock Data**   The following information pertains to the securities of Service Corporation International.

Table 3   Service Corporation International Securities

Common Shares Outstanding	2,870,446
Estimated Float	1,700,000 Shares
Warrants Outstanding	140,000
Estimated Institutional Ownership	301,700 Shares
Total Market Value	$48.8 Million
Average Daily Trading Volume—1980	4,400 Shares
Where Stock Is Traded	NYSE
Ticker Symbol	"SRV"
WSJ Listing	"SvCpInt"
Margin Status	Marginable
Options Available	None
Liquidity Factor	Moderate
Volatility Factor	Moderate
Investment Quality	Business-Risk
S&P Rating	B+

Table 4 Service Corporation International Annual Operating Results and Projections ($000)

Year Ending April 30	1976	1977	1978	1979	1980	1981 Estimate
Revenues	$80,125	$80,807	$89,116	$97,591	$115,009	$ 129,500
Pre-tax income	9,624	7,827	9,174	11,154	13,704	16,100
Net income	4,973	3,989	4,693	5,879	7,404	8,700
E.P.S. (fully diluted)	1.30	1.12	1.30	1.61	2.02	2.70–2.80
Dividends per share	0.115	0.20	0.24	0.28	0.34	0.44
Pre-tax profit margins	12.0%	9.7%	10.3%	11.4%	11.9%	12.4%

Table 5 Service Corporation International Condensed Balance Sheet ($000) April 30, 1980 (Audited)

Assets		Liabilities and Equity	
Current assets		**Current liabilities**	$ 15,315
Cash	$ 8,659	Long-Term Debt	68,450
Receivables	13,587	Deferred Items	15,281
Inventories	7,322		
Other	1,155		
Total current assets	$ 30,723	**Total liabilities**	$ 99,046
Fixed assets (net)	100,001		
Goodwill	13,526	Equity	62,816
Other	17,612		
Total assets	$ 161,862	**Total liabilities and equity**	$ 161,862

Questions for Analysis and Discussion

1. Discuss the description of "The Company" (pp. 209–211) as an example of technical description (see pp. 41–43 of Chapter Two for the characteristics of a technical description).

2. Using Hinton's headings, write a table of contents for this report. Then discuss the logic of its structure. What function does each section serve? Why has Hinton ordered his report in this way? How does each section relate to those that precede or follow it?

3. Examine and discuss the relation between the generalizations Hinton presents in the "Summary and Conclusions" section and the facts that support these generalizations. What kind of evidence does he offer? How detailed is his evidence? How convincing is it?

4. Imagine that you have $5,000 to invest. On the basis of this report and considering your expectations about the financial aspects of the next five years of your life, would you consider investing in Service Corporation International? Why or why not?

Application

To see how the same data can be used to communicate differing impressions, select a major publicly traded corporation and compare its annual report to its shareholders with the 10K report. (Both reports may be available in your university library.) Then obtain a financial analysis of the same company prepared by an independent stock or brokerage firm. Compare the impressions conveyed by each report and discuss the relative credibility of each. Pay particular attention to the use of selective omission of information in the annual report to shareholders. On the basis of your investigation, write a brief evaluative memo to your instructor, explaining your research and reporting your conclusions concerning the differing pictures presented by each form of report.

GLOSSARY

Abstract words   General or non-specific language. Unlike concrete language, which has particular or limited references (see *Concrete words*), abstract language refers to general concepts or intangible objects—for example, *freedom, sadness.*

Analogy   Comparison of an unfamiliar object, person, event, or concept with a familiar one, in order to show parallels or similarities and so increase understanding of the unfamiliar.

Analysis   The separation of an object, concept, or group into its component parts. For example, an analysis of the U.S. government by function reveals the executive, legislative, and judicial branches.

Classification   The arrangement of things, individuals, events, or ideas into general groups or classes by virtue of shared qualities, attributes, or characteristics. For example, individuals can be classified into groups according to age, sex, education, or occupation.

Coherence   The clear and logical relation of each part of the writing to all other parts of the writing. Writing is said to have coherence if the reader can easily follow the development and progression of sentences, paragraphs, and sections. Coherence is achieved through logical patterns of organization (see *Organization*) and effective transitional devices which indicate that organization (see *Transitions*).

Comparison and Contrast   Showing how two subjects—persons, places, objects, events, or concepts—are similar to and different from each other.

Concrete words   Specific and precise language. Concrete words refer to observable or measurable qualities of particular people, places, objects, or events. Unlike abstract language (see *Abstract words*), concrete language often establishes sensory details and exact measurements. For example: The tree growing outside my window is a three-foot, slender birch sapling.

Connotation The associations and overtones some words can imply in addition to their literal or dictionary meanings. For example, *thin, slim,* and *skinny* have roughly the same meaning, or denotation, but carry very different emotional suggestions, or connotations (see *Denotation*).

Coordination A sentence pattern that makes ideas of equal importance grammatically equal. For example: I tested the first batch, and he tested the second. (See *Sentence structure.*)

Denotation The stipulated or agreed-upon meaning of a word; the dictionary definition of a word, exclusive of its emotional overtones or associations (see *Connotation*).

Description Presenting the physical appearance, actions, or attributes of an object or concept. In technical writing, description gives an objective, verifiable account of its subject, often making use of narration when dealing with processes or objects that change (see *Narration*). Description is often a component of longer forms of technical writing (see Chapter 2).

Ellipsis A sentence pattern which omits words that are implied by the context. For example: The first group was accepted, the second denied. (See *Sentence structure.*)

Etymology The history of a word, its roots, and the way its meaning has changed and developed through time.

Exemplification Illustrating a general concept by describing a specific instance or example.

Explication Interpreting or making clear; especially used to explain the meaning of a key word in a logical definition (see *Logical definition*).

Figurative language The use of vivid, dramatic, or forceful expressions, not in their literal sense, but to reveal unexpected connections or associations. For examples of particular kinds of figurative language see *Metaphor, Simile,* and *Personification.*

Format The arrangement of words and graphics on a page and of pages in a document; the layout or physical appearance of a document. Format includes the spacing, margins, typeface, and headings of each page, as well as the division of the document into sections or chapters.

Graphics Visual aids or pictorial representations of a piece or set of information. Some of the most commonly used graphic forms are tables, line graphs, bar charts (or histograms), pie charts, pictographs, flow charts, diagrams, drawings, and photographs.

Inversion A sentence pattern that reverses the usual or expected order of words. For example: Totally unexpected were these results. (See *Sentence structure.*)

Jargon Highly specialized technical language, often understood only by members of a particular profession. Sometimes jargon is the most economical form of expression, but frequently it is a vague, abstract, inflated, or unnecessarily complicated way of expressing a simple idea.

Logical definition (or Formal or Sentence definition) The definition of a *term* by placing it into the general *class* to which it belongs and then separating it from all other members of that class by virtue of its *differentiae,* or distinguishing characteristics. For example: The automobile *[term]* is a means of transportation *[class]* that carries two to six persons, travels on four wheels over roads, and has an internal combustion engine *[differentiae]*.

Metaphor   The identification of two unlike things to imply a similarity between them. For example: A malignant tumor is a ticking time bomb.

Narration   Presenting a sequence of events, usually in chronological order. Narration tells a story, describing how something happens, often in the order in which it happens. Narration is frequently used in technical writing to describe processes or incidents.

Negation (or Elimination)   Explaining what something is *not* in order to clarify what it is; eliminating a particular person, event, object, or concept from all others with which it might be confused.

Objectivity   A detached and impersonal quality in writing. Writing is said to be objective if it deals with quantifiable observations and logical arguments without being colored by the emotions, prejudices, or biases of the writer (see *Subjectivity*).

Operational definition   Directing the reader to a time or place where the phenomenon, process, or object to which the term refers can be observed; describing the conditions under which the phenomenon defined comes into being. For example: A hurricane is the kind of storm that occurred in New Orleans in September 1965.

Organization   The way a writer arranges material in order to fulfill the purpose for which he or she is writing. The organization of a piece of writing can be simple or complex, and can follow a form unique to the writer or conform to a general pattern of organization. Some of these patterns include:

- *Chronological* (following a sequence in time)
- *Spatial* (following the physical layout of a place or object)
- *Classification* (establishing general categories)
- *Analysis* (dividing a subject into component parts)
- *Induction* (arguing from particular to general)
- *Deduction* (arguing from general to particular)

Parallelism   A pattern that balances related ideas within sentences, paragraphs, or series of paragraphs by using similarly structured words, phrases, or clauses. For example: Before contacting the subjects and beginning the experiment, it is necessary to read all instructions, to measure all ingredients, and to prepare the experimental apparatus. (See *Sentence structure.*)

Parenthesis (or Parenthetical element)   A qualifying word or phrase inserted into a sentence which interrupts the structure of that sentence. This word or phrase can be set off from the rest of the sentence by dashes, commas, or parentheses. For example: The samples, all collected by different groups, were analyzed immediately.

Personification Giving human qualities to nonhuman objects or abstract concepts. For example: The computer is a personal servant in today's world. (See *Figurative language.*)

Persuasion (or Argumentation) Convincing the reader to accept the writer's point of view. Although persuasion can appeal to the reader's emotions, in technical writing persuasion usually appeals to the reader's reason. Rational persuasion uses logical arguments, coupled with supporting facts, data, and examples. In technical writing, persuasion is particularly important in proposals.

Readability The ease with which a reader can comprehend a piece of writing; the adaptation of the language of a document to the document's purpose and audience.

Sentence structure (or Syntax) The order of words in a sentence, to establish meaning or to create effects. For examples of specific sentence structures, see *Coordination, Subordination, Parallelism, Parenthesis, Inversion,* and *Ellipsis.*

Simile The comparison of two unlike things, using *like* or *as* to imply the similarity. For example: The rocket went up like a geyser. (See *Figurative language.*)

Stipulatory definition Specifying or limiting the meaning of a term for the purposes of a particular occasion.

Subordination A sentence pattern that makes one statement grammatically dependent on and logically related to another. For example: Although it appears harmless, this organism can be lethal. (See *Sentence structure.*)

Style The overall effect of all the individual choices a writer makes; the "voice" of the writer. This includes the writer's word choice, sentence structure, paragraph development, use of detail, organization, and tone.

Subjectivity The insertion of personal attitudes into the writing. Writing is said to be subjective if it offers the personal opinions and emotional responses of the writer instead of conveying objective information (see *Objectivity*).

Tone The attitude of the writer toward his or her material and audience. Tone is indicated by ideas, words, and sentence structure. The tone of a piece of writing can range from formal to informal, serious to flippant, pompous to casual, pleading to restrained, sincere to ironic.

Transitions Words, phrases, sentences, or paragraphs that show the connections between one idea, topic, or statement and another. These devices help make writing more coherent (see *Coherence*). Transitions can indicate:

- *Relation* (therefore, however, consequently, likewise, similarly, in contrast, on the other hand, for example)
- *Time* (before, after, later, soon, until, during)
- *Place* (here, there, above, below)
- *Sequence* (additionally, furthermore, then, first, second, finally)

Unity   Treating all the ideas and language in a piece of writing in such a way that they are clearly relevant to its purpose. Writing is said to have unity if all elements cohere (see *Coherence*) and if there are no unnecessary digressions in content or shifts in form and style.

Word choice (or Diction)   The selection of the most suitable word or phrase for a particular context, taking into account the word's meaning (see *Denotation*), its associations (see *Connotation*), and its appropriateness to both audience and purpose.

Acknowledgments (continued from p. iv)

Chapter 2   *Pen*—From *How Things Work.* Copyright © 1979 by Encyclopaedia Britannica, Inc. Reprinted by permission. *High Clouds*—L. R. Koenig and C. Schutz, *A Temperate-Zone Cyclonic-Storm Model,* United States Air Force Project Rand, 1974, pp. 27–30. *Soil Stabilization: Materials*—Section 2.01 of "Soil Stabilization" from Beaver Creek Village Hall Specifications, Volume 1. Reprinted by permission of Hall Goodhue Haisley and Barker. *PHILIPS 2001 Specifications*—Reprinted by permission of Philips Information Systems. *Environmental Protection: Celanese Technology . . .*—"Celanese Technology Is a Step Ahead in Wastewater Treatment," *Celanese World,* Volume 5, Number 4, 1980. Copyright © 1980 Celanese Corporation. Reprinted by permission. *The Celanese Anaerobic . . .*—"The Celanese Anaerobic Wastewater Treatment Process." Reprinted by permission of Celanese Corporation. *Oil Shale Processing.* An Assessment of Oil Shale Technologies, U. S. Government Printing Office, 1980. *Processing Information in a Data Base*—Reprinted by permission from IMS/VS Version 1 Application Programming: Designing and Coding, SH 20-9026. Copyright © 1981, 1980, 1978, 1977, 1976, 1975, 1974 by International Business Machines Corporation.

Chapter 3   *How to Change Oil*—"How to Change Oil" from *Machinery Maintenance.* Copyright © 1981 Deere & Company. Reprinted by permission. *Beef Stew*—Armed Forces' *Recipe Service.* U. S. Government Printing Office, 1980. *Owner's Handbook*—McLane Owner's Handbook, Lawn Edger and Trimmer. Reprinted by permission of McLane Manufacturing Inc. *Taste*—Section 211, "Taste" reprinted by permission from *Standard Methods for the Examination of Water and Wastewater,* 14th Edition, 1976. American Public Health Association, Washington, D.C. *Carcinogens: What the Employer Must Do*—From *Carcinogens,* U. S. Department of Labor, Occupational Safety and Health Administration. U. S. Government Printing Office, 1975. *How to Plan . . .*—"How to plan the entire product support program." Copyright © 1980 Ken Cook Co. Reprinted by permission. *Here Is What . . .*—"Here Is What Mechanics Want in Maintenance Manuals" by Clyde Cheney. Copyright © 1980 by Technical Publishing, A Division of Dun-Donnelley Publishing Corp., A Company of the Dun & Bradstreet Corporation. All rights reserved. Reprinted with permission from the November 13, 1980 issue of *Plant Engineering* Magazine.

Chapter 4   *Table of Contents and Title Page from Proposal to the Electric Power Research Institute; Proposals for Solid Waste Management Planning in Fresno County*—Reprinted by permission of SCS Engineers, Long Beach, CA. *Memorandum: Shaw Hygrometer*—Memorandum from John R. Van Surdam, "Shaw Hygrometer for Monitoring Cargocaire Moistures." Reprinted by permission. *Floor Topping Project for X Plant*—"Floor Topping Project for X Plant." Reprinted by permission of Badische Corporation. *RERC Proposal*—Proposal to Community Development Corporation and Standard Proposal Terms and Conditions reprinted by permission of Real Estate Research Corporation, Chicago, Illinois. *Request for Proposal; Proposal*—TDG Proposal for Development Program for Boulder Junction Independent School District reprinted by permission of Michael D. Murphy.—*Grant Application: Florida's Endangered Papilio Species*—Research Proposal, "Critical Habitat Determination and Recovery Plan Development for Florida's Endangered *Papilio* Species" prepared for Florida Game and Fresh Water Fish Commission. Reprinted by permission.

Chapter 5 *Monthly Progress Report*—"Four Corners Monthly Progress Report, Particulate Removal Project." Reprinted by permission of Arizona Public Service Company. *Trip Report*—Trip report reprinted by permission of Vector Electronic Company. *The New Media . . .* —"The New Media and the Demand for Studio Production Facilities" by J. N. Dertouzos and K. E. Thorpe. Copyright © 1981 The Rand Corporation. Reprinted by permission. *Dwarf Sumac . . .* —"Dwarf Sumac as Winter Bird Food" by Jean W. Graber and Pamela M. Powers, *The American Midland Naturalist,* April, 1981. Reprinted by permission. *Financial Analysis . . .* —Financial Analysis of Service Corporation International. Reprinted by permission of Underwood, Neuhaus & Co.

INDEX